Everyday Nature

Everyday Nature

KNOWLEDGE OF
THE NATURAL WORLD IN
COLONIAL NEW YORK

SARA S. GRONIM

Rutgers University Press
New Brunswick, New Jersey, and London

Library of Congress Cataloging-in-Publication Data

Gronim, Sara Stidstone.
    Everyday nature : knowledge of the natural world in colonial New York / Sara S. Gronim.
       p. cm.
    Includes bibliographical references and index.
    ISBN-13: 978–0-8135–4024–5 (hardcover : alk.)
    1. New York (State)—History—Colonial period, ca. 1600–1775. 2. New York (State)—Intellectual life—17th century. 3. New York (State)—Intellectual life—18th century. 4. Nature and civilization—New York (State)—History. 5. Philosophy of nature—New York (State)—History. 6. Natural history—New York (State) 7. Science—New York (State)—History. 8. Landscape—New York (State)—History. 9. Human ecology—New York (State)—History. 10. Great Britain—Colonies—America—History. I. Title.
F122.G8 2007
974.7'02—dc22   2006021857

A British Cataloging-in-Publication record for this book is available from the British Library.

Manufactured in the United States of America

For Dennis

Always, and for everything

Contents

Acknowledgments

It is with deep pleasure that I thank the many people who have been so generous with their help over the course of this project.

My first thanks go to Thomas P. Slaughter, whose wisdom, good humor, and patience saw the first version of this study through to daylight as a dissertation. His insights were enormously helpful at key moments when I was lost in a thicket of evidence. I deeply appreciate his continued advice and counsel.

I have been extremely fortunate in the graduate education available to me, first at Brooklyn College and then at Rutgers University. Both institutions support superb history faculty, people who are committed to both scholarship and their students. I would particularly like to thank Philip Gallagher, Edwin G. Burrows, Paula Fichtner, Paul G. E. Clemens, Jan Lewis, Philip Greven, Gerald Grob, and Jennifer Jones. The members of the history departments, first at New Jersey Institute of Technology and now at C.W. Post campus, Long Island University, have been great colleagues. Talking through problems and concerns with friends has been enormously helpful. For their conversations I thank Serena Zabin, Robert Churchill, Robert Shaffer, Jaclyn Miller, Peter Messer, Greg Knopf, Lucia McMahon, Jenny Brier, Judith Van Buskirk, Julie Livingston, Conevery Bolton Valencius, and Michael Sappol. John Murrin and Patricia Bonomi have offered particularly helpful advice and encouragement.

The work of archivists and librarians is essential to any historian's work. I deeply appreciate the assistance of people at the following institutions: New York Public Library, New-York Historical Society, New York State Archives, American Antiquarian Society, American Philosophical

Society, Public Records Office (Great Britain), Linnaean Society of London, Edinburgh University Library, Alexander Library of Rutgers University, the Malloch Rare Book Room at the New York Academy of Medicine, Bobst Library of New York University, the Pierpont Morgan Library, the Botany Library of the Natural History Museum (London), and the National Library of Sweden.

Audiences at the following conferences offered perceptive comments at various stages of development: the American Historical Association, the Columbia Seminar in Early American History, the Conference on New York City History, the History and Sociology of Science Seminar at the University of Pennsylvania, and the Omohundro Institute for Early American History and Culture. Portions of this book appeared in articles published in *Explorations in Early American Culture, The William and Mary Quarterly,* and *The Bulletin of the History of Medicine.* I thank them for permission to reprint material here.

Several people provided concrete assistance with this book itself. My daughter Sara E. (Sasha) Gronim read the entire manuscript with intelligence and insight. Her comments on what was interesting and readable, and what was not, were enormously helpful. Caroline Rabinovitch was an efficient and effective fact-checker. I deeply appreciate Lyman Lyons's insightful and collaborative approach to copyediting. Jim O'Brien was expert with the index. Ruth Homrighaus also provided assistance at a key moment. I thank them all. It has been a pleasure to work with Rutgers University Press. Audra Wolfe as the acquiring editor offered excellent revision advice. Kendra Wood and Marilyn Campbell shepherded it through the production process.

Lastly, my family—Dennis, Spike, and Sasha—have been enthusiastic and loving supporters of my work. I am so very lucky to have them.

Everyday Nature

Introduction

NATURE AND KNOWLEDGE IN A
COLONIAL ATLANTIC WORLD

On a summer's day in 1705, a Dutch farmer from Claverack in the Hudson Valley of New York stumbled upon a curious tooth. Revealed by erosion in the bank of a stream that flowed toward the Hudson River, it was a "great prodigious Tooth," weighing a full four and three-quarters pounds. The farmer sold the tooth for a gill of rum to one Peter van Bruggen, a local member of New York's colonial assembly. Van Bruggen then brought it with him when he traveled downriver on assembly business to the port town of New York, where New York's royal governor, Lord Cornbury, expressed a desire for it. Believing it "worth nothing," Van Bruggen cheerfully gave it to the governor as a gift. Lord Cornbury in turn sent it to London to the Royal Society for its growing natural history collections. Cornbury also asked a minor official in Albany to look around for similar specimens. And indeed, later that summer men dug up the bones of a corpse thirty feet long (which sadly crumbled upon being touched), and the next summer people found more. Local River Indians, drawn to the scene of the digging, told these colonists that the bones confirmed a story told to them by their fathers, a story of a "monstrous person" long ago who had waded into the Hudson and caught fish with his bare hands. Another tooth surfaced from this dig, and this one an enterprising Dutch trader took with him on a trip to Massachusetts, where he showed it to the governor there. "I am perfectly of the opinion," wrote the Massachusetts governor, Joseph Dudley, that this was "a human body, for whom the flood only could prepare a funeral."[1]

The story of the Claverack bones shows a range of ways in which people responded to encounters with the natural world on the cusp of the eighteenth century. If local Indians interpreted these bones as part of their own claims to superior knowledge of this particular place, local farmers apparently found them mere curiosities, easily traded or given away. The erudite Lord Cornbury, in New York for a six-year stint as governor, by contrast saw them as a contribution to a great project of collecting and cataloguing the world's natural phenomena, a contribution to formal science. And Governor Dudley knew them to be an affirmation of Christian sacred history, of the great Flood that only Noah, his family, and their animals survived. Indians' history, natural history, sacred history, or a random curiosity "worth nothing"—people knew the significance of these bones in their own ways. And they did so without much exerting themselves to prove the others wrong.

In the seventeenth century, people who colonized Long Island, the Hudson River Valley, and the islands like Manhattan that lay at the mouth of the Hudson brought with them the complex of beliefs and practices common among ordinary people in early modern Europe. These beliefs and practices required only modest adaptation to the environment of northeastern North America, although colonists altered some practices to accommodate the new natural world they encountered. Because they brought with them technologies like sawmills, plow agriculture, and livestock raising that transplanted relatively easily, and because they were secure in their assumptions about how the natural world worked, they were only modestly interested in the knowledge of local Indians.

However, over the course of the eighteenth century, a variety of new beliefs and practices circulated in the wider Atlantic world. A broad movement that we in retrospect call the scientific revolution advanced a number of claims about the nature of the natural world. A study of colonial New York, however, shows that advocates of new understandings were only partially successful in persuading ordinary people of these new assertions. Some of these were adopted readily by many New Yorkers, others roused the interest of only a few, and still others provoked downright hostility. Rarely, if ever, did New Yorkers adopt or reject these new understandings based on simple debates about or demonstrations of their truth. Rather, social conflicts, religious allegiances, and political positions all strongly shaped the degree to which New Yorkers would change their minds about how they understood the natural world. Social relations of knowledge—what one assumed one could know for oneself and whom one could trust as reliable sources of what one didn't know—were as important as the content of the knowledge itself.

𝒩ew York was one of the many places around the rim of the Atlantic where colonists from Europe encountered and interacted with a new natural world. This particular region was first colonized by Europeans in the 1620s as an outpost of an expanding Dutch commercial empire because the Hudson River flowing from the interior of the North American continent offered access to the lucrative fur trade. The farmer who first found the giant tooth was a second- or third-generation descendent of those first settlers from the Netherlands and the adjacent territories of northwestern Europe. If the fur trade had been the initial lure, by the turn of the eighteenth century most people in the colony, like our farmer, made their living from agriculture. Even in the port town of New York at the mouth of the Hudson, shopkeepers, ship captains, craftsmen, and their wives kept gardens, chickens, and hogs.

Unlike many other colonies, New York had not attracted huge waves of colonists. By the turn of the eighteenth century, a census counted a mere eighteen thousand people (including slaves but excluding Indians). About five thousand people crowded into the port at the tip of Manhattan and a few hundred were upriver in Albany, and the rest were scattered along the hundred and fifty miles of the Hudson between New York and Albany or along the hundred-mile length of Long Island. Coastal Algonquians and Hudson River Indians, while considerably diminished by disease, continued to live in and among these colonists, while to the west the Five Nations of Iroquois remained diplomatically powerful and their territory intact. Consequently, interactions between Indians and colonists were common.

Our tooth-finding farmer was Dutch, but his governor was English, for the colony had been seized by England in 1664. As Van Bruggen's Assembly membership shows, colonists in New York, led by migrants from New England who had settled on Long Island, checked the power of the royal governor with their own legislative assembly. Our farmer, however, would not have voted because he was a tenant, not a landowner. His landlord was one of a handful of families who had accumulated huge grants of land in the seventeenth century and who dominated local politics, along with the merchants who increasingly vended New York's agricultural products. But even if a few dominated the commerce and politics of the colony, such a sparsely populated colony did not yet yield great riches for anyone. There were no leisured gentry, no closeted scholars in colonial New York. In 1705 New York was a modestly settled colony, one in which virtually everyone worked with plants and livestock, coped with earth and weather, sailed the Hudson River and Long Island Sound, and did their own healing work and household work. The natural world was not in pictures or in books, but was the fabric of their everyday lives.[2]

Part 1 of this book describes the natural world of these colonists from about the middle of the seventeenth century up through the first decades of the eighteenth. Those who settled New Netherland and then New York in the seventeenth century brought with them beliefs and practices about the natural world common among the socially and economically middling sorts in Europe. Chapter 1 discusses the land itself and colonists' efforts to reshape and re-imagine it through agriculture, land surveying, and mapping. Chapter 2 describes what colonists believed and did concerning the everyday: why bodies ailed and how one healed them, why patterns in the night sky changed and how that mattered, the what and wherefore of wind and weather. In chapter 3, colonists grapple with the anomalous, departures from the everyday like witches, epidemics, comets, and malformations of everything from kittens to radishes.

While colonists certainly responded to the material natural world that they found before them, nonetheless they paid only modest attention to what was peculiarly local. To some degree they needed to adapt only modestly because they had come to an area of North America whose environment was not so very different from the places they had left behind. Stocked with experiences in agriculture, fishing, and craft production that transferred relatively easily to the temperate climate of Long Island and the Hudson Valley, colonists were to some degree insulated from the necessity of mastering the minutiae of the local natural world. And they needed to attend only modestly to the local natural world because classical learning taught them that the natural world operated everywhere the same, and the Bible taught them that it was everywhere a text of God. God had created an intimate universe, one whose patterns could be understood by ordinary people. And indeed, those who left us records of their observations and experiences in personal letters, government reports, almanacs, and travelers' accounts were generally literate but not learned, for few with university educations or experiences among the upper classes sought their fortunes in such an out-of-the-way colony. By and large, these colonists brought with them everyday assumptions about the way the natural world worked and about how they themselves should work within it.

Over the course of the eighteenth century, this knowledge of the natural world held by ordinary people was challenged by newly developing sciences, and these challenges are the subject of part 2. However familiar some of these new assertions may be to us—assertions like the earth revolving around the sun, lightning as a manifestation of electricity, anatomical dissection as a useful tool in developing medical knowledge—these eighteenth-century changes in the knowledge of the natural world were embedded in social forms that were different from ours. In the seventeenth and eighteenth centuries generally there was no "science" or "technology" or "medicine" that inhabited realms separate from ordinary life, either institutionally or conceptually, as they would come to do in the nineteenth and twentieth centuries. Rather, in New York as elsewhere, innovations in the knowledge of the natural world were embedded in larger cultural shifts.

Consequently, part 2 is organized by these realms of cultural change. Chapter 4 examines various forms of improvement related to a rising faith in progress: agricultural innovation, inoculation for smallpox, and the organization of improvement societies. In chapter 5, the rise in refinement in personal taste, deportment, and sentiment was mirrored in new ways of approaching botany, electrical phenomena, and medical education. Chapter 6 explains that as reason came to be more explicitly cultivated, some tried to draw new lines between reason and superstition, lines that were intended to alter beliefs about witches, magic, and astronomy. In the final chapter, we return to landscape, both natural and human, to look at changes in beliefs and practices concerning cartography and race. Changes in beliefs about and practices in the natural world were part of these cultural developments, from claims about improvements in agricultural methods, to the elegant collecting of exotic plants in gardens, to the reasonable mastery of the orbits of comets, and to the disinterested display of geographic knowledge on maps.[3]

*B*ut even if these new sciences became steadily more familiar, their acceptance was by no means straightforward. In New York, as elsewhere in the eighteenth century, agriculture diversified, consumption of elegant goods rose, and institutions that cultivated reason like schools and newspapers were established. And yet, despite the development of New York from a rather isolated colonial outpost in the seventeenth century to one that seemed much more like provincial Britain by the third quarter of the eighteenth century, the colony retained much that was peculiar to itself. As in other colonial places, the institutions of formal religion became more pervasive, as churches were built and ministers hired. New York was particularly religiously diverse, with eleven distinct Protestant denominations active in the colony by the middle of the eighteenth century (as well as one congregation of Jews). While toleration was official policy, actual governors' policies sometimes favored one church over others, and some churches were themselves rent by internal dissension. Religion was thus often a source of controversy.

As elsewhere, the pace of commerce rose in the eighteenth century, and consequently New York ships increasingly sailed all the trade routes of the Atlantic and Caribbean, and prosperity within the colony climbed. But so too did commercial rivalries among New Yorkers sharpen, and disputes over taxation, access to land, and trade policies were heated and bitter. Dutch and English colonists were joined by people of German, French, Scottish, and Scots-Irish origins, who formed sometimes thin-skinned and clannish enclaves. New Yorkers' treatment of those they held in slavery grew harsher, and slaves' resistance added to the social turmoil of the colony. If large land-owners and merchants in New York now rose to real wealth, they did not become a unified gentry whose claim to political authority was a lofty devotion to the public good. Rather, they too fought constantly with one another, fights in which their own material stakes were evident to everyone. Ethnically and religiously diverse, New York was a chronically contentious, particularly disputatious place. And these disputes could take as their subject, among many other things, the nature of the natural world.

If the history of the scientific revolution was once written as the product of the minds of a few brilliant men, historians in the past thirty years have shown how important the participation of many more ordinary people—instrument makers, navigators, gardeners, doctors, and hundreds of collectors and experimenters who never became famous—was to the development of this shift in understanding nature. Moreover, historians now understand how crucial colonial places were to the scientific revolution. Breaking the insularity of Europe, where the erudite once believed that classical learning contained all there was to know, the newly encountered star patterns, plants, climates, animals, and people forced European savants to develop entirely new systems of thought in order to encompass this plethora of information. As importantly, people in colonial places themselves—not only European colonists, but Africans, Asians, and Native Americans—contributed their expertise to the growing catalogue of the world's natural phenomena. The history of the scientific revolution, then, is a global one, with many people in many places reworking their understandings of the natural world.[4]

But the scientific revolution was more than simply new assertions about the factual nature of the natural world. It also transformed social relations of knowledge. For example, a series of astronomer-mathematicians from Copernicus to Newton replaced a model of the universe in which the sun and planets circled around the earth with one in which the earth was itself a planet that circled, like all the others, around the sun. Not only was this a radically different assertion about what nature was really like, but it also was a novel claim about who really knew nature and how they knew. If ordinary people watched the paths of celestial bodies through the sky and thus could affirm for themselves the common understanding that they circled around the earth, by the end of the seventeenth century increasingly accurate predictions about the movements of celestial bodies like planets had confirmed this sun-centered model. But these predictions required specialized instruments to perform observations and sophisticated mathematics to calculate the paths of such celestial objects. As with the paths of planet, only the erudite could lay claim to legitimate knowledge of even objects available to ordinary experience, like the identity of a plant or the cause of an illness. Others, now asked often to disbelieve the evidence of their own senses, had to take this knowledge on trust.[5]

New Yorkers' skepticism about these new assertions concerning the natural world stemmed not so much from simple disputes over matters of fact, but from these challenges to familiar social relations of knowledge. In the long run, matters of truth are always functions of social relations. People are moved to change their minds—or to refuse to do so—depending upon the social resonance of what they are being asked to change. Who advocates or adopts the change is as important as what the change itself is. Social relations characteristic of New York shaped what was and was not incorporated from the array of new practices and beliefs that increasingly circulated in the Atlantic world. Whether it was the adoption of practices like hemp cultivation or smallpox inoculation, or ideas like Linnaean taxonomy or Copernican astronomy, the integration of innovation was accomplished to the degree that it was congruent with other realms of New Yorkers' lives. Whose claim to enhanced social authority was strengthened if the new claim was accepted? Whose beliefs and practices would now be denigrated as foolish or superstitious? Whose assertion about the nature of God would be affirmed or denied?

For people in the early modern world, innovations in ideas and practices in the natural world had implications beyond the practical, for the natural world was a realm of God, a model for human society, a theater for the demonstration of social allegiances, and a site for the performance of political power. Moreover, early modern people shared an orientation to the everyday and to the solidity of their own experience. In this economy of knowledge, the literate but not learned conceded very little authority to the erudite or the socially elevated. Innovations in understandings and practices that could be incorporated into familiar social relations and familiar material practices stood a good chance of being accepted. Those innovations, however, that bolstered some people's social power, religious claims, or political goals at the expense of others were as often as not vehemently resisted.

In this, New York was probably not much different from other places around the Atlantic rim in the seventeenth and eighteenth century. Indeed, I would expect that the experiences of people in Charleston and its hinterlands, or Newport or Dublin or Kingston and their surrounding areas, would follow a broadly similar pattern of slow and selective adoption, although one that might vary in its details. In all these places, the relative isolation of the seventeenth century shifted in the eighteenth century as the pace of commerce quickened, more people traveled as traders, missionaries, soldiers, officials, and settlers, and as the volume of print, both books and newspapers, brought news of innovation everywhere. In all these places, then, the everyday understandings of ordinary people were challenged by new assertions about the nature of the natural world. Everywhere this new knowledge was disseminated by newspapers and almanacs, in conversations in coffeehouses and around tea tables, and through demonstrations of everything from electricity to anatomy. If the pace and extent of these changes occurred variably in South Carolina, Rhode Island, Ireland, and Jamaica, nonetheless these changes were common to the Atlantic world as a whole. In concentrating on just one of these places around the rim of the Atlantic, I am attempting a fine-grained study of one particular place in order to think about this question of why some people might change their minds when presented with new claims about knowledge, and why some might not.

But if in some ways New York was like every other place in the early modern Atlantic world, there are reasons why it lends itself particularly well to an examination of middling people's relationship to the scientific revolution. For one, no one lived in New York who is particularly important to the formal history of the scientific revolution. No Benjamin Franklin arose to fame in New York, no Linnaeus or Newton ever lived there. There is no one, therefore, who so dominates the story as to overshadow the beliefs and behaviors of more ordinary people. Nor did New Yorkers establish an institution like the Royal Society in London or Yale College in Connecticut that would come to epitomize the state of learning in the area and so command a preponderance of attention. New York was an ethnically and religiously heterogeneous colony, commercially expansive and politically contentious, one in which no single person or group claimed center stage.

Its very heterogeneity and contentiousness is the other reason why New York is such a revealing place in which to examine the dissemination of new ideas about the natural world. In watching them quarrel over all these things, we can watch ordinary people, the literate but not learned of the early modern world, grapple with what the scientific revolution meant. The embrace of inoculation for smallpox, the acceptance of Lewis Evans's maps, the acknowledgment of Copernican cosmology, the interpretation of Franklin's experiments with electricity, the significance of plants just outside their own doors: all were subjects for debate and dissent. In all of these disputes, innovations could threaten what ordinary people did in their everyday lives and elevate some people's social authority while debasing that of others.

We who live on the far side of modernity have become so accustomed to associating innovation with progress and to granting some people the authority to tell us what is true (think how often a newscaster begins a story with "Scientists say . . ." or "Doctors have found . . .") that we can be unaware that these epistemological assumptions are not universal. Colonists in New York, by contrast, shared understandings of the natural world—why the body ailed, what made crops thrive, how patterns of the stars in the night sky changed—in which there were only modest roles for people with particularly accentuated knowledge. There were some roles that required special skills and knowledge—surveyors, navigators, midwives, doctors, almanac makers—but these people's knowledge was merely an accentuation of what everyone knew, not knowledge of a dramatically different kind. Moreover, the difficulties of the tangible natural world—epidemics, drought, storms—rendered people more skeptical of innovation than are we, who are generally so protected from laboring in the natural world by our social networks and our material infrastructure. Everyone did healing work, everyone struggled with crops, everyone used the skies as a timekeeper and season-keeper. Even if literate sources like the Bible and classical learning lent many of these understandings their coherence, nonetheless the proof of this knowledge was in material experience, in what one knew for oneself. The scientific revolution challenged much more than simply commonly accepted information; it was a revolution in how people knew the natural world. In a colony like New York, a place chronically short of social trust, it was a slow and difficult one.

PART I

New York as a
Colonial Outpost,
1650–1720

CHAPTER 1

Landscape

❧✦❧

We want the Musick of the Cuckoo still,
Yet in his stead we have our Whip-poor-Will.
Here's no Magpyes nor Rooks that Winged are,
Yet we have Birds that are in Britain Rare.
Indeed, our Singing Birds, of the same kind
With British Birds have shorter Notes, we find.
  Daniel Leeds, The American Almanack For . . . 1710

*I*n 1670 Daniel Denton concluded a description of the colony of New York by saying, "[I]f I have err'd, it is principally in not giving it its due commendation." Denton had arrived in the colony as a young man in 1644 with his father, a minister who took up a parish in the settlement of Hempstead on Long Island. In 1656 Denton became a local landowner himself. When Denton wrote his pamphlet, *A Brief Description of New York*, the colony had been undergoing settlement for almost fifty years, yet was still rather thinly settled. Thus far the port town of New York on the tip of Manhattan Island wasn't much more than a village, nor were there many more than a few hundred colonists up the Hudson River in the only other significant town, Albany. While more than a dozen small settlements were spread out on Long Island, there were only a handful along the Hudson between New York City and Albany.

"How many poor people in the world would think themselves happy," wrote Denton, if they had "an Acre or two of Land, whilst here is hundreds, nay thousands of Acres, that would invite inhabitants." Denton described land "of a very good soyle, and very natural for all sorts of English Grain; which they sowe and have very good increase of, besides all other Fruits and Herbs common in England, as also Tobacco, Hemp, Flax, Pumpkins, Melons, &c." And more than that: New York, he wrote, abounded with songbirds, fields of wild strawberries, "green silken Frogs," swift "Christal streams," and wild roses so beautiful that "you may behold Nature contending with Art, and striving to equal, if not excel many Gardens in England."

13

Nor did prospective colonists need to fear local Indians, for "there is now but few upon [Long Island], and those few no ways hurtful but rather serviceable to the English." To Denton's readers, New York must have seemed like England remade as Eden.[1]

Ten years later a traveler told another story. Jaspar Dankers, along with a traveling companion Peter Sluyter, had taken ship in Amsterdam in 1679, bound for North America to find a site for the settlement of their religious sect, the Labadists. Dankers and Sluyter docked in New York City, looked over western Long Island, and traveled up the Hudson to Albany. While Dankers found much to admire—fine peaches, productive sawmills and gristmills, and a waterfall on the Mohawk River that made him marvel at such a "great manifestation of God's power and sovereignty"—nonetheless he also found much that made him uneasy. Dankers found New York to be "a wild worldly world" because "most all the people who go there to live, or who are born there, partake somewhat of the nature of the country, that is, peculiar to the land where they live." Amidst the fruitful agriculture and those people of whom he approved, Dankers also recorded illness, poverty, and sin. Especially as he journeyed up the Hudson, where colonists lived "somewhat nearer the Indians," Dankers found that people were "wild and untamed, reckless, unrestrained, haughty and more addicted to misusing the blessed name of God and to cursing and swearing." If Denton's portrait was of a colony where a harmonious natural world sustained a fruitful social order, Dankers's New York was still too much a wilderness.[2]

Why were these accounts so different? Certainly, the two men wrote for different purposes, for Denton had a stake in presenting the colony to best advantage, and Dankers sent information to his fellow believers so they could choose their settlement site wisely (they chose Maryland). Denton had lived on Long Island since boyhood and hence was writing about a familiar landscape; Dankers, by contrast, saw with a stranger's eyes. Yet the ambiguity lies with us too, for we vacillate between conflicting interpretations of colonists' encounter with the natural world in North America. Sometimes we tell a story in which people coming from Europe become transformed into Americans by their interactions with American nature and by what they learn from American Indians, so that the natural world became their birthright and a source of their distinctiveness. And sometimes we tell a story in which those same people were culturally blind, wanting only to remake the local landscape into the image of Europe, and so were hostile or indifferent to what was particular to it, and to what the people who were already there knew. Like many such dualisms, both contain an element of truth, and neither is entirely satisfactory.

$\mathcal{P}$eople who came to colonize New Netherland and then New York indeed sought to transform the landscape into something like the one they had left behind. They brought with them familiar practices like European farming, surveying land in order to divide it up into private property, and mapping it so as to bring it within the scope of European learning. And yet they could not transform the land altogether. For one thing, the local landscape and ecology had their own peculiarities that resisted being remade into a mere replica of somewhere else. For another, even if colonists were consistently concerned with their material sustenance, and preferably their own prosperity, there were real limits to how thoroughly they could exploit the natural world. Bearing with them a technology of iron and wood, migrants to North America nonetheless remained at the mercy of the elements and under the limitations of human and animal muscle. Lastly, the social landscape of the colony never replicated the societies they had left behind. That, perhaps, mattered most.

One of the limits to colonists' mastery of local nature was simply knowing what was there. If we look at a map of the area dating from the 1650s, Nicolaes Visscher's *Novi Belgii Novaeque Angliae nec non partis Virginiae* (New Netherland and New England besides part of Virginia), what is striking is the extent of empty space, reflecting the mapmaker's uncertainty about what was in the interior (illustration 1). In the seventeenth century, the Dutch were the premier cartographers of Europe, producing beautifully engraved maps based on information gathered from sea captains carrying on the Netherlands' far-flung commerce. The Visscher map was the most widely circulated map of New Netherland, reprinted and pirated well into the eighteenth century (with "Niew Jorck" merely added to the engraving after the English took the colony in 1664.) But navigators rarely ventured beyond the shoreline and on the Visscher map "New Netherland" was a general area rather than a clearly demarcated place. The shorelines and the easily navigable portions of rivers are firmly drawn, but the rivers beyond that just run generally up into the interior. The Saint Lawrence River, known from French accounts, runs dramatically, if inexactly, across the top of the map, but the mapmaker was apparently aware of only one of the great interior lakes. Probably he intended to represent Lake Champlain, which is on what is now the New York-Vermont border, but he knew of it only generally and so put it northeast of the Connecticut River, into which it apparently

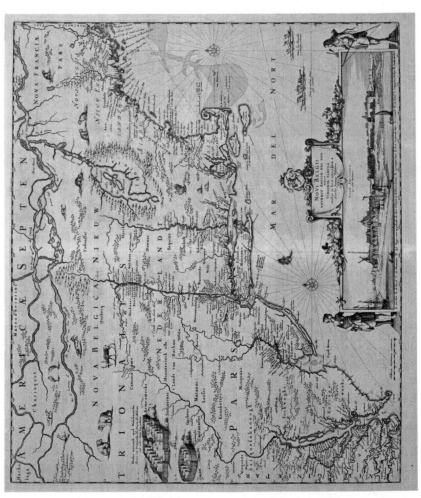

ILLUSTRATION 1. Nicolaes Visscher, *Novi Belgii Novaeque Angliae nec non partis Virginiae tabula multis in locis emendata* (1651–1656). One of the earliest maps of the area, it was based on mariners' actual observations of the coast, but could only sketch out the interior, as little of that had been seen by Europeans. *(Reproduced with permission of the I. N. Phelps Stokes Collection, Miriam and Ira D. Wallach Division of Art, Prints, and Photographs, The New York Public Library, Astor, Lenox and Tilden Foundations.)*

emptied. On the landscape itself, he scattered drawings of trees and hills and wildlife, what a European audience by the seventeenth century expected of the North American interior, a place that was a wilderness.[3]

Few of the people who colonized the area, however, would have seen such a map, for most were of too modest means. Rather, what they saw was the land itself. Long Island ran east to west roughly a hundred miles, and was about twenty miles at its widest. Left by the retreat of the glaciers of the last ice age, its southern shore sloped gently into the Atlantic, and indeed was somewhat indistinct, as sandy barrier islands just offshore changed their configurations with every winter storm and much of the shoreline was lost in tidal marshes. The northern shore was higher and more wooded, with deep harbors opening out onto the sound between Long Island and Connecticut. The islands in the bay at the mouth of the Hudson—Manhattan, Staten Island, and several smaller ones—were varied woodland and marshland, ridge and stream. The Hudson flowed almost straight south from its source some three hundred miles in the Adirondack Mountains, and was wide and deep enough for ships to sail a hundred miles into the interior of the continent; shallow draft boats could sail all the way to Albany. On its west bank rose low mountains, the Catskills. The east bank rose steeply in many places too, but just inland were forests on deep, rich soil with swift, clear streams. While there were few places in North America that had an ecology that was more like that of northwestern Europe from which most colonists came, nevertheless the landscape hardly looked like Europe.

Yet the landscape that prospective colonists saw as they arrived in the seventeenth century was not a wilderness either, nor had it been for hundreds of years, perhaps even a thousand. Over the course of centuries, migrants from the interior of the continent had moved toward the Atlantic coast, slowly adapting their corn, beans, and squash agriculture to new ecological niches, and mastering the local patterns of the game and fish they encountered. Algonquian-speakers had spread along the Atlantic coast from what is now Virginia to Massachusetts, Iroquoian-speakers stretched from Lake Champlain to Lake Erie, and Mohicans and others lived in the Hudson Valley. All of them had transformed their local landscapes in roughly similar ways: villages sat amidst acres of corn fields, trails knit together villages and hunting territories, fishing weirs dotted shorelines, and much of the forest was cleared of its underbrush by periodic deliberate fires. Small camps near shellfish beds or prime hunting grounds dotted the rest of a group's territory, and sites of old and abandoned villages were in various stages of being reclaimed by woods. Small game and wild berries flourished in the low brush of abandoned fields. Well before any Europeans arrived, Indians had remade the landscapes of eastern North America.[4]

The people who migrated to the area from Europe in the seventeenth century could not have colonized the region had it been otherwise, nor would they have wanted to. When the Dutch West India Company first promoted settlement in the area in the 1620s, they did so because local Indians, particularly the Mohawks and the Mohicans up the Hudson, could supply them with valuable beaver pelts and other furs. Indeed, on Visscher's map, what dots the interior are the names of Indian groups, and depictions of two versions of Indian villages lay along the left side of the map. If the physical attributes of the shores and rivers mattered to navigators, the distribution of Indian people was what really mattered about the interior. Even colonists who were not themselves engaged in the fur trade availed themselves of an Indian-altered landscape. Early settlers more often than not erected their farms on old Indian sites, taking advantage of the cleared land, and travelers by land followed Indian trails.[5]

In the 1680s, for example, several dozen families established the little villages of Nyack, Haverstraw, and Tappan Landing on the lower Hudson at the heads of Indian trails into the interior. And in 1715, when the surveyor John Reading was employed to seek potential mines and fertile land in northwestern New Jersey, he followed Indian trails, stayed in Indian villages, and recorded potentially useful sites in his journal with Indian-named landmarks. The Indian-altered landscape made European colonization possible, just as commercial relations with Indians made colonization worthwhile.[6]

Even if colonists initially settled following Indian patterns, they worked steadily to transform the landscape into one much like the pastoral landscapes of rural Europe. Dutch colonists brought the market gardening, orchards, and grain growing of the Netherlands to the first settlements on and around Manhattan and up the Hudson at Albany. In the 1640s New Englanders began migrating to Long Island where they introduced English forms of agriculture (which historically had borrowed much from the Dutch), particularly wheat farming and dairy and cattle raising. In the 1660s a group of Dutch families established Schenectady, a farm community on the rich alluvial flats beside the Mohawk River just northwest of Albany. To do so, they used diking and drainage techniques so valuable in the Netherlands. Other Dutch, joined in the 1680s by Huguenot refugees, farmed the creek bottoms of the west bank of the Hudson. Indeed, by the time Denton and Dankers wrote their accounts, trade in foodstuffs and livestock produced in the colony was beginning to outstrip the fur trade in volume and importance.[7]

Yet they did not entirely replicate the farms of northwestern Europe. For one thing, clearing land for crops took precedence over the labor of clearing land for meadows and fencing them. Consequently livestock were largely left to forage freely. On Long Island, scattered natural grasslands lent themselves to the collective grazing of sheep, cows, and horses, and towns sometimes hired a herder. But everywhere swine ran wild in the woods, and horses and cattle often did so too. By the late seventeenth century, free-ranging livestock had become a problem. With animals invading gardens and trampling crops, town after town passed regulations requiring fencing, marks on livestock to forestall disputes over ownership, rings in the noses of hogs to prevent them from rooting up grasses, and yokes around their necks to prevent them from wriggling through fences. Rather than lessen their dependence on husbandry in the absence of the labor to create fenced pastures, then, colonists improvised different practices.[8]

Moreover, colonists learned from coastal Algonquians and Hudson River Indians the technique of burning woodlands. Each fall Indians burned out the underbrush of large expanses of woodlands to facilitate the tracking of animals during the upcoming winter, and colonists quickly saw the advantages to this. In October of 1716, the ship on which the Irishman John Fontaine traveled moored off Sandy Hook, New Jersey, just south of Manhattan, unable to navigate safely through the thick smoke "occasioned by the burning of the woods" onshore. Lewis Morris, who farmed north of Manhattan in what is now the Bronx, explained to a recent arrival that colonists burned the underbrush for different reasons than did Indians. Indians do it, Morris wrote, "to render the tracks of deer, and other wild creatures they hunt, more conspicuous." Colonists, on the other hand, did it for two reasons: to make a firebreak to keep their farms from being burned when Indians fired the woods, and to burn away dead leaves and grass so that new plants could emerge more easily in the spring. This makes "a fortnights difference in the Spring," he assured the newcomer, to a farmer's ability to turn his livestock loose to forage for themselves.[9]

As colonists appropriated an Indian-prepared landscape and adopted some local ways of managing it, so too did Indians adopt agricultural practices from colonists. Especially on Long Island, where colonists settled most densely in the seventeenth century, Montauketts and others began to raise livestock and grow wheat. Jasper Dankers described a Nyack village on the western end of Long Island during his visit in 1679–1680. The Nyack grew corn and beans, as Dankers had expected, but also had hogs and chickens, and an orchard of peach trees. By the early eighteenth century, even Indians up the Hudson had adopted some aspects of European farming, as when the sachem of Schaghticoke asked Albany colonists for help with their annual plowing in 1711. Not that this proceeded without conflict. The town records throughout the colony are filled with disputes over livestock belonging to colonists or Indians that destroyed the other's crops, or over the boundaries of grazing rights. Thus the agricultural practices of colonists and Indians over the first several generations of colonization became somewhat more alike, neither entirely European nor entirely American.[10]

Plants and animals imported from Europe were the mainstay of colonial agriculture, but some colonists showed modest interest in local plants and animals. An early settler, Adriaen van der Donck, wrote rapturous descriptions of the colony that included a list of over three dozen wild plants. "[I]t is not to be doubted," he thought, that many of them would prove to have value in healing. Similarly, Denton described the variety of native fruits and trees, and named seven plants he thought might have healing potential. But except for using trees for building and firewood, settlers in New York developed little knowledge of local plants. Like the strawberries Denton lauded, they gathered berries akin to those they knew in Europe, and to a modest degree they adopted corn-growing from the Indians. On rare occasion a colonist shipped a plant back to Europe as an exotic. In 1656 and 1657, for example, Jeremias van Rensselaer, who supervised his family's land near Albany, sent tubs with sassafras trees to his uncle in the Netherlands. German refugees from the Palatinate, deposited in an uncultivated area of the Hudson Valley by a governor who had brought them to New York to produce naval stores, recalled their suffering by pointing out that "had it not been for the Charity of the Indians, who shewed them where to gather some eatable roots and herbs, must inevitably have perished every soul of them." But this was an exceedingly rare example of colonists eating wild plants. Coastal Algonquians, River Indians, and Iroquois had of necessity and through long experience learned the minutiae of the local natural world. By and large, colonists preferred the grains, vegetables, and orchard fruits they brought with them. And the very point of colonization was to extract the region's commodities for profit, weaving the territory into a world of trade, and wild plants offered little for that.[11]

Colonists showed more interest in local animals. Early travelers often commented on the abundance of wildlife. Access to a plentiful supply of beaver pelts, of course, drove the colonization of the area from the beginning. And the Dutch in particular, who had historically first accumulated wealth as a people through herring fishing, marveled at the masses of fish in local waters. Jacob Steendam, who lived in the colony for eight years, published a poem titled "The Praise of New Netherland" that included these verses:

> The lamprey, eel and sunfish, and the white
> And yellow perch, which grace your covers dight;
> And shad and striped bass, not scarce, but quite
> Innumerable.

The poet went on to name seventeen other distinct species of fish before concluding:

> There's not a pool or tiny water trace,
> Where swarm not myriads of the finny race,
> Easily taken.

Oysters abounded in coastal shallows and apparently everyone harvested them. But while fishing was a widespread skill in northwestern Europe, hunting was not, although Dankers observed Dutch colonists occasionally shooting wildfowl or deer. The one wildlife skill colonists sought to learn from Indians was whaling. Beginning in the 1660s, companies of Long Island settlers, accompanied by Montauketts and Shinnecocks, began to put to sea to catch whales. But neither beaver trapping in the north nor shooting or trapping deer and birds in nearby woods seemed to have interested these settlers. They came with farm and craft skills from Europe, and farmers and craftsmen they remained. Rather, deer and wildfowl were touted as a boon because Indians hunted them and sold them to colonists "at an easie rate." Local wildlife did become part of colonists' diet, but that was largely because they could simply buy it.[12]

Despite these accommodations to local conditions and the adoption of some local practices and foods, colonists still looked around them and saw a wilderness. In the 1630s an Amsterdam merchant, Kiliaen van Rensselaer, bought thousands of acres of land near Albany. Van Rensselaer never came to the colony himself, but after his death the family sent out a series of relatives to manage the estate. In 1669 Jeremias van Rensselaer, the man who had sent his uncle the sassafras trees, wrote his mother of the loneliness of being "in this distant and strange country," and worried to his brother that the family was risking too much in "this heathenish" place. "[How] slow our Conveighance is," fretted an English governor in 1670, "like the production of Ellephants." He thanked a correspondent for long-awaited letters, adding "if you did but know what darkness wee live, as if wee had as well crost Lethe, as the Athlantiq occean." Isolated from Europe in a rough, unfamiliar landscape, "wilderness" was both a description of a place and a metaphor for the absence of home.[13]

Colonists repeatedly compared what they saw with what they had left behind, and many found it difficult to love the local landscape. I would never prefer, wrote John Miller, chaplain to the soldiers in the fort on Manhattan in the 1690s, "the wild Indian country before our English meadows and closes, much less our gardens when in the most flourishing estate." Daniel Leeds, who settled in New Jersey in the 1680s, commented that "Creatures in this Country move sooner to Generation and Corruption than in England; so in the Vegetable Kingdom the Fruit Trees, especially of Orchards, bear fruit younger, and the Trees are sometimes of smaller growth, and shorter lived, than the same sorts in England." Still, slowly, familiarity softened some of the strangeness of the landscape. Dankers reported that some colonists had named natural features after memories of home, like a hill called "Boterberg [Butter Hill] . . . because it is like the rolls of butter which the farmers in Holland take to market." Leeds also wrote the verse that opened this chapter, comparing the songbirds of England and America. The ambivalence of the poem is characteristic of a colonist in its simultaneous eye for the possibilities of an unfamiliar place and the mournful loss of home.[14]

Yet wilderness was not merely metaphorical, for the immediate landscape of New York offered genuine material threats, two of which were wolves and poisonous snakes. Wolves, gone for generations from the Netherlands and England, still lingered in remote areas like the Scottish Highlands and loomed large in the imagination of migrants to the New World. The embodiment of savagery and a bodily threat to the husbandry that stocked a civilized landscape, wolves were a clear enemy everywhere in the colony. Consequently, town after town from the east end of Long Island to Albany and Schenectady paid bounties, often to Indians, for wolf pelts. Rattlesnakes were also common everywhere in the colony, and they clearly stunned colonists. We have many snakes "as long as 8, 10, and 12 feet," wrote Johannes Megapolensis from Rensselaerwyck in the 1640s. "Whoever is bitten by them," cautioned Adriaen van der Donck in 1650, "runs great danger of his life." Wolves and rattlesnakes made it difficult for colonists to think that they had entered an entirely benign landscape.[15]

Another way in which the local natural world confounded expectations was its climate. Prior to their colonial expansion, Europeans had expected that places in the same bands of latitude would have the same climate. Yet New the region was hotter in summer and colder in winter than the northwestern Europe from which most colonists came, despite lying on the latitude of Portugal. One year Jeremias van Rensselaer wrote home to Holland that the winter had been so severe that several neighbors had their hands or feet frozen. "If this be your Spring," a local wit had a visitor from Barbados ask a New Yorker, "pray Sir what is your winter?" Observers occasionally offered a local explanation for the incongruous climate. New York, explained Charles Wooley, who lived there in the late 1670s, is "full of Lakes and great Vallies, which receptacles are the Nurseries, Forges and Bellows of the Air, which they first suck in and contract, then discharge and ventilate with a fiercer dilatation." Leeds bemoaned the harshness of the climate in another verse:

> The Climate here from England differs wide
> In Weather, Change, and something else beside,
> For we are scarcely here a year about
> But North-West winds do find us out,
> and blows us sharply one against another,
> Friend against Friend, and brother against brother.

Colonists repeatedly compared what they saw with what they had left behind, torn between their eagerness to see the local natural world as full of productive potential and their experiences with its disconcertingly alien aspects.[16]

*L*eeds's verse on the climate indicates that colonists recognized that wilderness was as much in themselves as it was in their landscape. As elsewhere in the seventeenth-century Atlantic world, magistrates and ministers strove to suppress a boisterous popular culture and bring their charges to a more thoroughly civilized decorum. Peter Stuyvesant, the Dutch governor from 1647 to 1664, worked tirelessly to control the drunkenness, brawling, petty theft, and sexual escapades that filled New Netherland's court records. In 1664 the Reverend Hermanus Blom of Wildwyck in the Hudson Valley asked local magistrates to ban "the public, sinful, and scandalous Bacchanalian days of Fastenseen [Shrove Tuesday], coming down from the heathens from their idol Bacchus, the God of wine and drunkenness, being also a leaven of popery." In 1678 English governor Edmund Andros fined seven Dutchmen and an Englishmen for disturbing the tranquility of the colony. Their offense had been erecting a "maypole" before the door of a newly married man. Evidently they deemed the man too old for his bride, for they decorated it with a withered tree, a straw wreath, and dried beaver testicles.[17]

In general, Andros bemoaned his difficulties turning New Yorkers into obedient subjects, attributing their recalcitrance to the nature of the original colonists, "adventurers," he wrote, who had acted "before any Regulation." Even if Dankers had attributed some of the wildness he saw in New York colonists to their proximity to Indians and to an uncultivated wilderness, he also had to admit that the sinfulness he saw had other origins. A Hackensack Algonquian pointed out to him (or perhaps Dankers put words in his mouth) that "we did not have so much sickness and death before the Christians came into the country, who have taught the people debauchery and excess." Many of the disputes that so racked public life in the colony were personal: petty thefts, adultery, insults, drunken brawling. However much Daniel Denton would have liked to persuade his readers that social harmony reigned in New York, court records, ministers' and governors' letters, and travelers' tales tell another story.[18]

If rowdiness was one source of social disorder in the colony, so too were antagonisms between groups. Certainly the colony was nothing if not ethnically diverse, with groups of Dutch, English, Scots, Germans, and French settling in the colony in its first hundred years. The change from Dutch to English law constrained women's rights to property, and impeded Dutch-speakers' opportunities. Adriaen Janse van Ilpendam, for example, was a notary in Albany whose Dutch literacy became less and less useful after the English conquest. He found himself growing old and marginalized, and eventually took his own life. Yet ethnicity was not always a clear-cut source of contentiousness. Amidst all this diversity, some moved easily. Jacob Melyn, brought as a child from Amsterdam in 1642 when his father bought land on Staten Island, moved with his family to New Haven colony in 1655, and married an English woman there. The couple moved back to New Netherland, settling across the Hudson, and then moved into New York City where Melyn operated a tannery under English governance. Finally, in the 1690s they moved to Boston where they lived out the rest of their lives. While some Dutch, then, accommodated to the new laws and government brought by the English, others retreated into a defensive hyper-Dutchness, and others, like many Germans, insulated themselves in rural areas where they would have little to do with outsiders. But while ethnicity certainly mattered, it was by no means the only source of chronic factiousness of New York.[19]

Many of the most persistent and corrosive disputes were over access to economic advantage. Sometimes the scale of these disputes was small, such as contentions over weights and measures when colonists sold each other bread or sold their wheat or other provisions to a long-distance trader. Sometimes they were over matters of public policy, such as disputes over whether the burden of taxation should fall more heavily on land or goods.

And as the colony slowly became more settled and more productive, opportunities to become wealthy rose. As they did, political alliances, particularly with the governors who rotated in and out of the colony, could mean the difference between thriving and failing.[20]

Under the English, governors granted land, gave contracts for services like supplying soldiers, appointed the judges who ruled on civil disputes like inheritances and contracts, and elevated the favored to a small circle of officials. Twenty-year-old Robert Livingston, born in Scotland but reared in the Netherlands, arrived in Albany in the winter of 1674–1675, bringing with him commercial skills and contacts. He married a young Albany widow, Alida Schuyler van Rensselaer, and the two of them developed a trading and land-owning partnership that eventually made them one of the wealthiest families in the colony. Crucial to their success were Robert Livingston's relationships with a succession of governors, for through them he got government contracts, land grants, and government positions. Other New Yorkers were similarly ambitious. Consequently, by the end of the seventeenth century, these New Yorkers were coalescing into factions that vied with each other for political and economic advantage. With considerable exasperation, in 1693 Governor Benjamin Fletcher described them as being as "much divided in their interest and affection as Christian and Turk."[21]

Another source of social dissension was the practice of slavery. Like other places in the Atlantic world, from the 1620s on colonists in New Netherland imported Africans as slave labor. As elsewhere, the boundaries between slaves and servants in New Netherland was initially somewhat fluid, with some slaves earning freedom, marrying, owning property, and becoming members of churches. By the time of the English takeover in 1664, there were more than seven hundred people in the colony held in slavery, with perhaps seventy-five people who were former slaves. However, the imposition of English law made indentured servitude more expensive, and thereafter the shift towards seeking slaves for labor accelerated. As colonists did so, they steadily tightened restrictions on slaves, with the General Assembly passing increasingly restrictive slave regulations beginning in 1681. These regulations were primarily intended to constrain slaves' freedom of movement. One hue and cry, for example, called for the recapture of one Jacob, a slave from Schenectady, who "Speakes good English and Dutch, and can read Dutch, he speakes good Maquase [Mohawk] and Mahikanders [Mohican]." Seldom held in groups of more that two or three by any one owner, slaves in New York shared in all the farm work, domestic work, craft work, and trading work that their owners did. People like Jacob, some of whom were born in the colony and some of whom had been seized in Africa, learned the social landscape very quickly. Because of the potential for escape to Indians, for example, a 1702 regulation mandated the death penalty for any slave found more than forty miles beyond Albany.[22]

While slave regulations were intended to render enslaved people quiescent and obedient, slaves were hardly so. In 1712 more than two dozen slaves in New York City rebelled against their fate: they armed themselves, set fire to a building, and then began shooting the people who arrived to put it out. They killed nine people and wounded seven before being put to flight by soldiers from the fort and local townspeople. Quickly surrounded, six of the rebels committed suicide; twenty-one others were executed. Colonists knew exactly why the slaves had risen up. "Some Negro Slaves here of ye Nations of Carmantee & Pappa," reported the Reverend John Sharpe, chaplain to the soldiers at the fort, "plotted to destroy all the White[s] in order to obtain their freedom." The leaders, identified by Sharpe as being from the Akan-speaking region of West Africa, were likely recent arrivals. Enslaving strangers would have been a familiar to them from home, but in the Akan region such slaves, if loyal, could be rewarded with land and the fruits of their labor, and their descendents would eventually be fully integrated into local kin networks. These new arrivals in New York were no doubt appalled at the hopelessness of their condition under the colony's version of slavery. This rebellion—along with the running away, recalcitrance, petty theft, and even murder that slaves did to resist and reshape the conditions under which they were forced to live—convinced a handful of colonists that slavery was a danger to New York. But the lure of such labor proved too strong. Slavery—with the discontent and social dissension it bred—would remain one of the many sources of the chronic contentions that so characterized New York.[23]

*H*ere, as in other colonies, one way to counter a disorderly social landscape and to promote tranquility was the transformation of the material landscape into private property. For ordinary people, title to their own land was desirable because it improved their chances for self-sufficiency and even prosperity. For colonial officials, land-owning heads of households could be held responsible for the good behavior of their dependents, and would be expected to uphold the formal institutions of law and government upon which their land titles depended. Granted, turning the land into private property was not the original impetus for colonization. The Dutch West India Company colonized the area solely because of access to the fur trade. It imagined a social hierarchy of colonists in which most would be either employees of the Company (soldiers, clerks, craftsmen), or farmers who would be tenants of "patroons," great landowners who would bear the costs of transporting colonists and developing roads and wharves. Only Kiliaen van Rensselaer actually took up a patroonship, however, claiming land that grew to hundreds of thousands of acres on the upper Hudson near Albany. More modest colonists pushed to acquire land of their own and in 1638 the West India Company acquiesced. When the English took over, the new governor required that people resurvey their property and reapply for title, but nonetheless English law consistently provided a route to acquire land as property. People's investment of money and labor in their own land would be the source of the fences, hedges, pastures, and fields of a properly domesticated landscape.[24]

And yet dividing up the land into property, whatever colonists' intentions, was not entirely successful either at domesticating the landscape or grounding social peace. While in the modern world we picture private property as something bounded by clear lines that are measured on the ground and specified on paper, in the early modern world property was often less distinct. While the people who came to colonize New Netherland and New York certainly understood the utility of defining property clearly both on the ground and on paper, several things undermined its actual practice in the colony. It was difficult to survey uncultivated, unmarked land, and few men in the colony had the mathematical skills to do more than estimate the acreage a plot contained. As importantly, social discord and political competition undermined the secure recognition of property.

Surveying seems in retrospect a rather straightforward task requiring a modest amount of skill. To carry out a survey, a surveyor began at a corner of a parcel of land and took a directional reading with his compass down one edge of the property. He then directed two men to lay out a measuring chain in a straight line along the edge of the property until they got to the next corner. At that corner the surveyor would take a compass bearing down the next edge of the parcel, and the chain-bearers would again measure the length of that edge. In such a way they would proceed around the perimeter of the piece of land. When they were finished, the surveyor would have the information necessary to represent the parcel on paper and calculate its area in acres. The clarity of the geometric representation of the parcel on paper and the precision of its dimensions expressed in degrees of the compass and the length of chain would seem to ensure that the property could ever after be identified, and the measurement of its perimeter, and thus its area, could be replicated by anyone with the requisite skill.

Nevertheless, measuring the perimeter of a plot of land by laying down a hundred-foot chain while encountering rocks, trees, underbrush, and uneven ground was necessarily imprecise. Compass readings varied with the natural unevenness of the earth's magnetic field, a variation that in any given place drifts slowly over time, and local iron deposits could also pull the needle slightly east or west. In 1723 Peter Fauconnier, a Huguenot who had come to New York in 1702 as secretary to Governor Cornbury, was hired by colonists on Staten Island to do a formal survey of their property lines. Although he carefully recorded the length of each side of every piece of property, when he sighted down the boundary lines with his compass he neglected to account for the local variation of the needle. Without this information, no surveyor in the future could duplicate his survey and thus confirm these boundary lines. Surveying, then, was at best an approximation, and the expense and difficulty meant that colonists as often as not dispensed with it altogether.[25]

Not surprisingly, the acreage listed on the resulting plat did not coincide reliably with the actual area of the property claimed on the ground. In addition, calculating irregular areas took greater mathematical skills than most men who moved to the colonies possessed until well into the eighteenth century. In 1681, for example, a Long Island minister described a quarrel between a widow and her neighbor over property lines. "Several surveyors, called in by the parties, did not succeed in removing the difficulties," he wrote, "but rather increased them; because they did not agree as to the proper understanding of the matter nor as to the measurements, and they gave different opinions." Most formal surveys merely offered a general estimate of the acreage contained in a plot, and no one seemed to regard a more precise calculation as particularly necessary. Some surveyors even appeared unable to write, as occasionally they signed their attestations with their marks rather than signatures. One report commented on some large grants as being "not laid out by exact measure of acres, but computed in the lump by miles." As was generally true in all the colonies, then, in New York surveying was often enough an approximation rather than a practice of precise measurement.[26]

What was crucial to the widespread acknowledgement that some parcel of land was in fact someone's private property was social agreement, not the mere act of surveying. When Fauconnier recorded his Staten Island surveys, he recorded that "We begun by the Stump agreed by all to have been one of the first former Mark[s]." Fauconnier duly marched around all the lot lines, following the courses people pointed out to him. In so doing, he was recording in measurements what was already secured socially, translating what was "agreed by all" into mathematical figures. His experience had been otherwise in the town of Rye on the mainland just north of Manhattan, for there local people ran off the surveying party's horses and the "malice and jealousy of the people" dogged their every step. Indeed, the colony's General Assembly proposed that in cases of dispute any "three Neighbours, Men of knowne intelligence" be empowered to walk the lines and render judgment. When neighbors agreed on local boundaries, the lines held; when no such agreement was forthcoming, a survey couldn't settle matters.[27]

Technical limitations and social conflicts were not the only problems that often left land title obscure. The process of acquiring legal title was expensive. In New Netherland and New York, colonial authorities never asserted that the land was "empty" and therefore colonists' for the taking, for Algonquians, Hudson River Indians, and Iroquois plainly occupied the land. By inserting themselves between colonists and Indians, officials tried to control the acquisition of property by colonists. A prospective purchaser was first supposed to get permission from colonial officials to purchase a particular parcel of land from its Indian owners and then get a signed deed from the Indians when the purchase was made. New York records are full of such warrants that name the Indian sellers and describe the land they sold. After securing Indian title, the purchaser had to get final title by having the parcel surveyed and then filing the plat and description with colonial authorities, all of which cost money. Only after all that was done was a government-sanctioned deed granted by the governor. Whatever had been Indians' understandings and uses of territorial space before the arrival of European colonists, Indians quickly came to understand colonists' idea of private land ownership. Their testimony about the boundaries of land they had sold or leased runs all through New York's colonial records, as when Esopus Indians walked the boundary of land they had sold in 1677. Getting title from Indians was expensive and time consuming; the subsequent fees to colonial officials for final title added to the cost. Small wonder, then, that many colonists did not complete the process.[28]

Moreover, vague boundaries left open the exact specifications of the land one claimed. Robert Livingston, for example, got formal warrants from Governor Thomas Dongan in 1683 and 1685 for twenty-six hundred acres of land he then bought from the Mohicans. Yet somehow when it was formally surveyed in 1714, the land Livingston claimed contained one hundred sixty thousand acres, one of the most egregious sleights-of-hand in colonial New York. The widespread reluctance generally among colonists in New York to pay first for purchases from Indians, then for formal surveys, and lastly for final fees to officials meant that the thirty-year gap between Livingston's purchase from the Mohicans and his formal survey of the land was not unusual. Such slow completion of all the steps in the process made it relatively easy to obscure how much land one was, in fact, claiming as one's own.[29]

If middling colonists sought their own prosperity through acquiring property in land, colonial officials sometimes used land in an attempt to create a different social order. As had the West India Company, several English governors late in the seventeenth century granted huge swaths of land to particular men in order to promote the emergence of the sort of landed gentry that held the preponderance of power in England. Robert Livingston was just one of a couple dozen men able to claim tens of thousands of acres in the late seventeenth century. These men were then expected to people their land with the kinds of industrious but deferential tenants that would both bring prosperity and social peace to the colony. A number of these grants, like Livingston's, were given the legal standing of "manors," whereby each was entitled to send a representative to the colony's General Assembly. Thus would such land grants create simultaneously a social and a political hierarchy.[30]

By the end of the seventeenth century, colonial officials posted to New York began to object to the land grants of their predecessors. Richard Coote, Lord Bellomont, who arrived in the colony as governor in 1698, complained in particular about "several most extravagant grants of land" made by his immediate predecessor, Benjamin Fletcher. Fletcher indeed had granted land lavishly, allowing local political allies like Frederick Philipse to lay claim to ninety thousand acres of prime land just up the Hudson from Manhattan. Few of these large grants had been much improved, nor were they attracting the sorts of deferential tenants that these aspiring men had hoped for. But as importantly, Bellomont and Fletcher were on opposing sides in London politics, and when Bellomont arrived in New York, Fletcher's local allies became his enemies. Land grants, then, while certainly a tool for promoting colonial development, were also a political weapon. Not only did Fletcher's grants empower men who would be Bellomont's political enemies, but they also (although Bellomont did not mention this in his complaints to London) left Bellomont that much less land with which to reward his own political supporters.[31]

To get the attention of his London superiors and recruit their assistance, Bellomont dispatched a map (illustration 2). Drawn by Augustin Graham, the colony's surveyor general, the map showed the most prominent grants of land in the colony and was intended to support Bellomont's contention that "this whole Province is given away to about thirty persons." Such a map would have been welcomed by London officials, for "send a map" was a frequent directive to their appointees out in the colonies. They well recognized that familiarity with a colony's geography could be useful in administering it from afar. Yet the dispatch of such maps was infrequent, for the men with skills to make them were rare in New York, as in other colonies, and neither London nor New York's General Assembly was willing to bear the cost of mapmaking.[32]

The map Bellomont sent, with its neat lines circumscribing the land grants he wanted to highlight, looked as if it represented mathematically determined property, for Graham put a scale of miles and a compass rose on his map, symbols that signified reliability. But the properties lining the Hudson were merely estimated because Graham had few formal surveys upon which to draw. Nonetheless, the clarity with which these grants were presented contrasted with ones whose extent and location could only be guessed at. North of the Mohawk River, along the southwest corner of Albany County, and along a line drawn to Lake Champlain (here labeled "Corlars Lake"), Bellomont had had Graham write in the terms of grants that were merely grand expanses. As the Board of Trade would report about one of them, it contained "about twenty four or thirty miles in length: its breadth we know not." Bellomont managed to get two of these poorly defined grants cancelled by the General Assembly before his sudden death in 1701, but the General Assembly resolutely refused to void others (which was hardly surprising, as some of their claimants sat as members). Nor did London officials take much action. Finally, in 1708 authorities in London ruled that no one in New York could henceforth be granted more than two thousand acres, and they left it at that. It was up to local officials to settle the relationship between access to property and social peace.[33]

Officials in London were less concerned with the ways in which changing land into property affected colonists' relationships with one another than they were with the effect of these undefined land grants on relations with the Iroquois, a confederacy of the Mohawks, the Oneidas, the Onandagas, the Cayugas, and the Senecas. All of the most vaguely worded grants lay within territory claimed by the easternmost nation of Iroquois, the Mohawks. Unlike coastal Algonquians and River Indians, the Iroquois had a political importance that transcended merely being a local source of land title or information. In the seventeenth century, the French in Canada were rivals of the Dutch and English for furs from the interior of the continent.

ILLUSTRATION 2. Augustin Graham, *A MAP of the PROvince of New Yorke in America and the Territorys Adjacent* (1698). A hand-drawn map, its object is to show that much of the desirable land in the colony had been given as huge grants to a few individuals. Their grants line the east bank of the Hudson River, and a few more, whose dimensions were vague, are noted near the Mohawk River and between the Hudson and Lake Champlain (Corlars Lake). *(Reproduced with permission of the Public Records Office, Great Britain.)*

The Iroquois, determined to be the primary brokers of the fur trade, warred against Indians allied with the French and thus the French themselves. Indeed, one of the primary tasks of governors dispatched to New York was to promote good relations with the Iroquois and thus lubricate this trade. By the end of the seventeenth century, France and England had expanded their rivalry to Europe and throughout their colonial empires. Consequently, the French in Canada, only two hundred miles north of Albany, became a military threat to New York. In February 1690, for example, the French and their Indian allies raided Schenectady, killing about sixty people and taking twenty-seven more as prisoners. New York governors had a mandate to persuade the Iroquois to ally themselves with the English, both as a conduit for trade and a military bulwark against the French.[34]

But even if London officials had some recognition of the significance of the Iroquois, by and large they were consumed with politics at home and on the continent of Europe. In 1701 Bellomont sent another map to London, this one drawn by Willem Wolfgang Römer, a military engineer dispatched by London to see to colonial fortifications (illustration 3). The map showed the territory of the Five Nations of Iroquois that ran westward from New York, and thus offered an implicit argument for New York's importance to England's imperial aspirations. Römer's map showed how easy it was for fur traders to move from the upper Great Lakes in the upper left corner of the map to New York on the map's right. Fur traders could canoe from Lake Ontario to Lake Oneida in the territory of the Iroquois, and from Lake Oneida only a short portage was required to move onto the Mohawk River and thence to Albany. Similar to Graham's map with which Bellomont had tried to marshal support for nullifying land grants, this map was part of an effort to procure more attention and more resources for the colony. But, as with Graham's map, this map had only a modest impact.[35]

Like Graham's map, this map purported to represent the geography of the colony as if it had been measured, for Römer, too, put a compass rose and a scale of miles on his map. But while Römer had a compass, his only measurement for distance was the movement of his body through the woods. And he himself had only traveled the route along the Mohawk River to Lake Oneida and from thence to the shores of Lake Ontario. On his map he exaggerated the dimensions of areas through which he had struggled, and underestimated the distances of places he knew only from hearsay. Maps like those sent to London by Bellomont offered more information about the geography of the interior than there had been on Visscher's map, for now colonists like Graham and travelers like Römer had seen some of this land

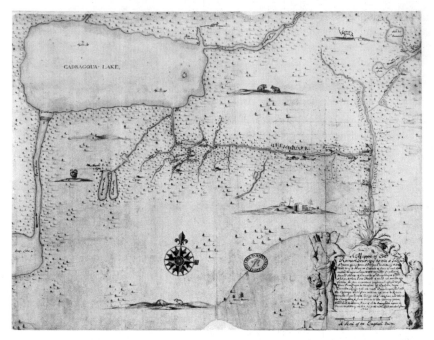

ILLUSTRATION 3. Willem Wolfgang Römer, *A Mapp of Coll. Romer his journey to the 5 Indian Nations* . . . (1700). The map makes it seem easy to travel between the Hudson River and Lake Ontario ("Cadragqua Lake"), and so emphasizes the route Indians around the Great Lakes could travel to Albany to trade furs. *(Reproduced with permission of the Public Records Office, Great Britain.)*

for themselves. But like survey plats, maps made in New York in the late seventeenth and early eighteenth century were largely approximations, for no one had the resources to do more.

If some people translated their experiences with the material landscape of New York onto the paper surface of a map, others did the reverse, trying to take information from a map and mark it on the actual ground. In 1719 the governments of New York and New Jersey appropriated money and appointed commissioners to run their common boundary, to fix the boundary that existed on paper to a line on the ground. New Jersey was represented by its surveyor general, James Alexander, a recently arrived university-educated Scot, and New York by Allane Jarrett, whom the acting governor described as "a person agreed on all hands to be the most capable of anyone in the Country." The two men set to work accompanied by an entourage of chain carriers, Indian guides, and assistant surveyors.[36]

As was common enough when boundaries in North America were decided in England and based on inadequate maps, local topography couldn't be made to fit the description on paper. When the Duke of York granted Sir George Carteret and Lord John Berkeley the territory of New Jersey, the area was defined on the west, south, and east by bodies of water. No such natural feature was available to mark its northern boundary, the line that separated it from New York. The Duke of York, most likely relying on the Visscher map, defined the division line between the colonies as a diagonal running northwest from a point at forty-one degrees of latitude on the Hudson River to a point on the Delaware River which was on "ye Northermost branch thereof which is in 41 Degrees and 40 minutes of Lat."[37]

To find one's latitude, a surveyor (or a navigator—the skill is the same) used an instrument like a quadrant to measure the sun's altitude, the angle the sun made with the horizon at noon. He then compared it with a table of solar altitudes for the days of the year to find his own position. The commissioners had no problem agreeing on the point on the Hudson that would anchor the easternmost end of the boundary line, for that point had in fact been agreed to long before. They then proceeded up the Delaware to determine the westernmost end. But there is no branch of the Delaware that is both "northermost" and intersected at forty-one degrees, forty minutes. The commissioners simply chose a branch of the Delaware by relying on Indian informants for assurance that this branch continued in a northerly direction (which it does.) They then set up their instruments in the Indian town of Kasheton (which was open enough for them to be able to measure the angle of the sun above the horizon). The two surveyors took sightings that persuaded them that they were at the correct latitude, and based on that agreed that here was the proper end of the dividing line.[38]

The agreement didn't hold. In the twenty years before the commissioners set out to draw the line, New York governors had made a number of large land grants in the area, grants that were imprecisely worded and were themselves un-surveyed. When Jarrett returned to New York, he abruptly declared that his instrument had been imperfect, that he now couldn't be sure his boundary line was correct, and that he needed a new instrument. "It is impossible for the Art of Man," fumed one New Jersey man, "to make an Instrument perfectly true and correct . . . if the Line be stay'd till One be certifyed to be so . . . it will be stay'd for ever." Yes, the money for the boundary commission was spent, land titles couldn't be secured in the area, and so Crown quitrents (a kind of land fee) couldn't be collected, admitted Peter Schuyler, the acting governor of New York, to the Lords of Trade. But, Schuyler added blandly, Jarrett couldn't be compelled to certify the boundary "without offering an injury to his conscience." With mapping as with surveying, social agreement was crucial, and if it was not forthcoming, no instrumentation or mathematical calculation could command consent.[39]

All this mattered little to most colonists in New York, for most people aside from navigators rarely, if ever, saw a map of any kind. Few seventeenth-century people lived in a "map immersed" world. For us, maps are everywhere: in the glove compartment, on placemats in tourist restaurants, at entrances to museums and amusement parks. We use them casually, thinking of them as tools for wayfaring. But travelers in the early modern world generally found their way not by consulting a map but by consulting other people. They relied on other people's physical experiences with an area, and through traveling made their own experiences. When militia captain Arent Schuyler, for example, reconnoitered the area between the Catskill Mountains and the Delaware River in 1694, he did not take a map but hired guides. In New York, much travel over any distance was by water, along Long Island Sound or up and down the Hudson, for waterways made natural roads; and so travelers merely put themselves in the hands of the captain. Otherwise, people simply asked directions.[40]

*B*y the early eighteenth century, after a hundred years of colonization, the landscape of New York had changed. A 1723 census counted just over forty thousand non-Indians (over six thousand of whom were slaves) in the L-shaped territory between Albany and the east end of Long Island. An early eighteenth-century traveler praised the houses in New York City, many of which were built of brick "of divers Coullers and laid in Checkers, being glazed [which] look very agreeable." By now New Yorkers shipped their agricultural produce to London and Amsterdam, to the West Indies, and up and down the Atlantic seaboard. Yet by no means was the landscape utterly transformed. Much land was not yet taken up as private property (or not yet openly claimed). Marshland and meadow on Long Island, ridges and bogs along the Hudson, woods and streams everywhere functioned as a commons for Indians and colonists alike. Towns still paid bounties for wolves, and free-ranging hogs plagued everyone. While it was an altered landscape, it still contained much that was undomesticated.[41]

It was not only colonists' material practices that mattered to changes in the landscape. The society they developed over the first hundred years of European settlement in the area was polyglot, dispersed, and contentious. Most of the people who came were modestly or little educated, and sought merely to carve out spaces for themselves where they with friends and kin could secure a decent living. Governors and a handful of the exceptionally ambitious, like the Van Rensselaers and the Livingstons, may have sought to develop a society that reflected their ideal social order, but they had only modest success. Neither a wilderness nor a mirror of Europe, the material landscape of New York reflected a society in which colonists of modest and varied origins invested their labor in some familiar ways (like establishing orchards and grain fields) but not in others (like fenced pastures), made do with mostly approximated surveys of their land, relished the abundance of fish and game, and wrestled with the threats from rattlesnakes and wolves.

While colonists had learned some things of importance from local Indians, their interests in what Indians knew about the natural world was limited: where a trail led, how to spear a whale, how to use a canoe. The records are full of links between Indians and land title, but are nearly silent on what Indians knew about specific local plants, how they read the night sky, or how they interpreted signs of changes in the weather. Part of the explanation for this is that much of Indians' knowledge about the local natural world had been painstakingly developed over hundreds of years, and was not easy to learn. We sometimes sentimentalize this as Indians' respect for nature, as if they were somehow nicer people than are we, but Indians mastered the specific peculiarities of the local natural world in minute detail because it mattered. Men who ignored the subtle sign of an impending blizzard and left on a hunting trip, women whose food stocks fell low because they had failed to exploit every bit of what they could grow and gather, died quickly. Indians knew the local natural world intensively because they had to, while colonists were largely insulated from their need to know what was local by the agricultural practices and other technologies they brought with them.

As we shall see, colonists brought with them their own understanding of how the natural world worked that they believed to be universal. They did not need to learn about healing or the stars from Indians because they knew what they needed to know already.

CHAPTER 2

Body and World

Each changing Season does its Poison bring,
Rheums chill the Winter, Agues blast the Spring.
Wet, Dry, Cold, Hot, at the appointed Hour,
All act subservient to the Tyrant's Pow'r.
And when obedient Nature knows his Will,
A Fly, a Grapestone, or a Hair can kill.
    Titan Leeds, The American Almanack For . . . 1728

"[*M*other] is taken with an extream fitt of Sickness," Robert Livingston Jr. wrote his brother urgently in 1716. "Since twelve of the Clock, the Symptoms seem to be worse than any she has had hitherto, she is siz'd with an extream pain all over her Body as if her blood was stop'd in its circulation." Alida Schuyler Livingston recovered, but a year later she was ill again. "I long to see you here," she wrote her husband, "for I am now getting worse . . . the phlegm are now so hard and sore . . ." In 1712 daughter Johanna journeyed to New Haven to tend her married sister Mary who was "verry ill whit hour [her] brest." Johanna stayed two months, then sadly wrote her parents that "sister Livingston [went] out this trobelsoum wourild the 8 instant."[1]

Thus do the letters of the Scots-Dutch Livingston family who lived in and around Albany describe the intermittent eruptions of illness in their family, and the fear and sorrow that illness caused. In the early modern world every episode of ill health was fraught with the danger of incapacity and death. Although colonists in northern colonies like New York would prove to be as healthy as people anywhere in the Atlantic world, sickness was a near-constant, if not of oneself then of kin and neighbors, and shifts between health and ill health a central preoccupation.

The literate but not learned people like the Livingstons who colonized places like New Netherland/New York brought with them from Europe a common understanding of how the body worked and how illness happened. They took it for granted that they had little to learn about such fundamentals from their engagements with either the local natural world or Indians who lived in the area. When Titan Leeds printed the poem above, he was repeating the obvious. The physical world was changeable, full of risk, one in which "a Hair can kill." Yet, "Wet, Dry, Cold, Hot," its fundamental nature was everywhere the same, in North America as it had been in Europe. For these early modern Europeans, both the human body and the surrounding natural world were constituted similarly, and the balance and flow of heat and cold, dryness and moisture were the sources of change in body and world alike. Indeed, body and world were intimately connected, for the skin was not a barrier but a permeable boundary, one that allowed conditions in the external world to affect the internal.

Colonists consistently claimed that the touchstone of reliability in these matters was "experience," their own bodily interactions with the material world and the evidence of their own senses. At the same time, colonists understood that what they learned from their experiences was to some degree provisional. The natural world was too much in flux and mankind was too imperfect to allow for eternal and incontrovertible knowledge. One person's experience often did not match another's, and hence there was room for variations in which practices actually worked, and colonists' expectations for mastery over the natural world were modest. Nonetheless, there was a coherent, consistent set of understandings through which colonists framed their experiences and thus rendered them comprehensible.

Because colonists shared a common body of beliefs and practices that were affirmed continually through their everyday experiences, they relied on the knowledge of others only modestly. The social relations of knowledge were such that virtually all everyday practices—like the farming and surveying in the last chapter—were to some degree within everyone's experiences. Everything from assessing the source of an illness to following the passage of stars in the night sky was within everyone's ken. In this, a colony like New York was different from Europe, for here there were no great cities, no courts, no universities in which the learned pronounced their interpretations of natural phenomena. In New York, the literate but not learned ordinary sort dominates the records. In such a society erudition was not much valued nor much missed. What people knew for themselves would do.

Their beliefs and practices were all affirmed in the print that circulated in the Atlantic world. In both the Netherlands and England in the seventeenth century there was an explosion of self-help books—books on agriculture, medicine, navigation, and surveying—all of which found a ready audience among the middling and better sorts. But importing such books

into the colonies was expensive. For example, books sold by the executors
of Gysbert van Imbrock, who died in the Hudson Valley in 1665, brought
significantly higher prices than those books had cost in Europe. Among
colonists in the seventeenth century, most owned no more than a Bible and
perhaps a primer, if they owned books at all. Ministers often had libraries of
some sort, and Governor Bellomont brought a library of over two hundred
volumes with him when he arrived in 1698, books he left to a New York City
minister. But the few lists of such collections that still exist show that their
titles were overwhelmingly devotional. In a world in which most relationships were still face to face, understandings of how the natural world worked
and how to work within it would have been primarily shared orally.[2]

Access to print expanded beginning in 1693, when William Bradford
set up the first printing press in New York. Bradford hired Daniel Leeds
to write the annual almanacs that were a mainstay of a printer's business.
In Europe almanacs were among the most common sort of reading matter
people owned other than devotional literature, with roughly a third of families purchasing one annually in England by the 1660s. No such figures exist
for the colonies, but the Livingstons ordered Leeds's almanac to sell in Albany, and Long Island shopkeepers offered them. Leeds offered local readers
much more than merely a monthly calendar. His calendars served colonists
as timekeepers, for they listed the times of sunrise and sunset, and the changing patterns of the sky at night. And they offered extensive advice for daily
life, recipes and advice on healing and agriculture. The information in his
almanacs, and those of his son Titan who succeeded him, was based on common understandings of the cosmos, and their practical advice derived from
the same print sources that circulated in Europe. Bradford himself drew on
those same sources when he reprinted a common English advice manual,
The Young Man's Companion, and then began to print New York's first agricultural and medical self-help guide, *The Husband-man's Guide*. By the end
of the seventeenth century, then, New Yorkers had their own printed advice,
but Bradford and Leeds tailored their advice very little to particularly local
circumstances. Rather, they assumed that both human bodies and the natural
world that surrounded them worked very much the same everywhere, and so
the old assumptions worked just fine in a new place.[3]

*T*he template for understanding the natural world was the human
body itself, the microcosm that mirrored the macrocosm of the great world.
Colonists understood their bodies to be composed of four humors, each with
its characteristic proportions of wetness or dryness and cold or heat, whose

balance and flow constituted health. Their bodies were intimately connected with the wider world through their permeable skin, allowing each season "its Poison bring." Shifts in the moisture and temperature of the surrounding air with the shifts in seasons and weather affected people, plants, and animals alike.

For colonists, as for Europeans elsewhere, these shifts were affected by changes in the positions of the sun, moon, and planets. Such effects could be predicted through astrology, the knowledge of a universe in which heavenly bodies were relatively close to earth and so easily exerted their effects. Daniel and Titan Leeds offered more than a reading of the sky as a timekeeper; they kept track of the movement of heavenly bodies through the zodiac, thus allowing their readers to connect body, the surrounding natural world, and sky. Even here, however, an area in which celestial movements were predicted mathematically, expectations for reliability were modest. As with surveying land, approximations would largely do.

While the Livingstons used words familiar to us, like blood and phlegm, they used them to refer to a set of beliefs about the physical composition of the human body that was considerably different from ours. "[W]hat else is the Body of Man," the minister George Keith asked rhetorically in 1694, "besides the four Elements of Fire, Air, Water and Earth?" From the ancient Greeks, early modern Europeans understood that all worldly matter, the human body included, was made up of varying proportions of these four fundamental elements. Each element, in turn, had a characteristic proportion of the qualities of heat and moisture: fire was hot and dry, air hot and moist, water cold and moist, and earth cold and dry. Circulating throughout the human body were four fluids or humors, each of which was a homologue of the four elements: blood (hot and moist), phlegm (cold and moist), yellow bile (hot and dry), and black bile (cold and dry.) When Alida Schuyler Livingston suffered pain that convulsed her whole body, her son attributed it to blood that could not circulate freely, perhaps because it had become cold and thickened, or perhaps because of some impediment. Similarly, when Anna van Rensselaer wrote from the Netherlands to her son about someone's unexpected death, she concluded, "It was nothing but melancholy [an excess of black bile] that was the matter with him." Although by the seventeenth century, concern for blood predominated over concern with the other humors, health was the appropriate circulation of these four humors in appropriate proportions in the body.[4]

Each person had a distinct "constitution," a specific configuration of these four humors. In general, a predominance of warmth and dryness made a person male and a predominance of coolness and moisture made a person female, although every individual had his or her own peculiar balance that was "health." This explained the basis for what were believed to be men's and women's distinctive mental capacities, with men's firm, dry minds better at sustained logical reasoning and women's relative moistness making their minds more impressionable. Individual temperament, too, was explained by the predominance of one humor or another, with people dominated by phlegm, for example, being dull and stolid, and choleric (yellow bile) ones quick tempered.[5]

Illness often stemmed from an excess of a humor, or a problem with its flow, as when Alida Livingston's blood was "stop'd in its circulation," or corrupted humors. Such illnesses were conceived to be derangements of the body as a whole, a conception that indeed matched people's most common experiences. In the early modern world, infectious diseases were ubiquitous and the source of perhaps the most common morbid condition, "fever." And fever, with the weakness and aches that accompany it, is an experience of the whole body, even when symptoms predominate in one part of the body. With the "hard and sore" phlegm in Alida Livingston's chest, her whole body suffered from the maldistribution and corruption of her humors, the misery of her dried-out, stuck phlegm. Although colonists recognized that some ailments— rotten teeth, broken bones, wounds, urinary stones—were located in specific body parts and called for localized treatments, most healing was directed at restoring the health of the body as a whole. Treatments were directed at purging the ailing person of excessive or corrupted humors and so restoring his or her proper balance and circulation. Evacuation could be through any opening in the body using any number of routes: letting blood, encouraging vomiting, defecation, urination, or sweating, or by drawing humors through the skin by blistering. On a difficult sea voyage in 1695, Robert Livingston recorded that one of the sailors became "raving mad, so that he was bled and put to sleep." With the balance of the sailor's body restored, tranquility followed.[6]

Colonists used an enormously eclectic variety of medicinal preparations in response to illness. Sometimes they sent for favored medicines from home, as Jeremias van Rensselaer did in 1664 when he asked his mother in the Netherlands to send a special salve for his wife's painful leg. When Bradford set up his print shop, he began to sell imported proprietary medicines (commercial compounds from a secret recipe), and other colonists imported them directly for their own use or for resale. Peter Stuyvesant asked for "medicinal seeds and plants" from the botanical garden at the University of Leyden, which the directors of the West India Company obligingly sent him. Mostly settlers made medicines from herbs they grew in their gardens and other substances close at hand. Daniel and Titan Leeds often included recipes for medicines in their almanacs, like this one offered in 1705: "The Herb Motherwort either in Syrup, Conserve, or otherwise, is reputed excellent to drive melancholy Vapours from the heart, and make a cheerful blyth soul." Bradford's *The Husband-man's Guide* suggested treatments for some twenty-five illnesses, using combinations of more than fifty ingredients. Colonists could swallow ingredients mixed in wine, honey, or fat to stimulate the evacuation of excessive or corrupted humors via diarrhea, vomiting, or sweat. They could ease pain by applying poultices made with softened grain or animal dung. They could heal wounds or burns with salves made with a base of beaten eggs or animal fat. Most of these ingredients came easily to hand in the kitchens, barns, and gardens of ordinary colonists.[7]

Little of this medicine was based on plants or practices that colonists found in the colony when they arrived. Even if colonists like Adriaen van der Donck had no doubt that local plants were of great value "principally because the Indians cure very dangerous and perilous wounds and sores by roots, leaves and other trifles," colonists in New Netherland and New York seemed to have made little effort to learn such information themselves. Daniel Leeds advised his almanac readers that eating dried Indian corn dried up "raw crude humors from the stomach" and that tobacco expelled tough phlegm from the lungs, stomach, and chest. But by the time he did so, knowledge of these American plants had long circulated in the wider Atlantic world. Overwhelmingly, the healing methods colonists brought with them from Europe were the mainstay of treatments. They had a panoply of healing substances already, and on these they primarily relied.[8]

The recipes in *The Husband-man's Guide* called for both plant ingredients and mineral substances like sulfur and mercury. These two sorts of ingredients stemmed from different formal medical traditions, the "Galenical" and the "Chymical." Galenical preparations were made largely from plants and were based on works associated with the second-century Roman physician Galen. Chymical medicines were of more recent origin, having been introduced into European medicine by the Swiss physician Paracelsus in the sixteenth century. Daniel Leeds expressed doubts about Paracelsus's innovations in his 1708 almanac:

> The Learned Physitians, such as were of Old,
> Galen and Hippocrates lie and mould;
> Now Paracelsus claims the curing part

And most Men practice the spagyrick art,
Yet Herbs when gathered in their proper seasons
More harmless physick makes, for divers reasons.

But whatever Leeds's doubts about the innovations Paracelsus had introduced, he himself was far from dogmatic in offering medical advice. For example, he prefaced his list of healing herbs in his 1695 almanac by saying "what here follows, is partly from experience, and I presume 'tis moderately true." Similarly, *The Husband-man's Guide* marked some recipes "Probatum est" (it is demonstrated), but left most unmarked, indicating that the author was surer of the reliability of some recipes than of others. Experience helped, but nothing worked all the time. This plethora of healing recipes and techniques reflected both the varieties of ailments and the limits of human knowledge. In healing, as in so much else, Daniel Leeds remarked, only God was infallible.[9]

Illness was so serious an experience, and healing so chancy, that prevention loomed large. In common with others in the Atlantic world, colonists in New Netherland and New York believed that the health of the body depended upon temperate living, moderation in all areas: activity, emotion, food, and drink. When Governor Henry Sloughter died shortly after entering a New York rent by turbulence and dissent, a group of local men attributed his death to the "perturbations, caused by conflicting passions and emotions." Food and drink were particularly important in keeping one's humors uncorrupted and in proper balance. Robert Livingston was not surprised when he got sick on his 1695 voyage from "the change of food and the hard, salted fish and oil." By contrast, Jaspar Dankers was relieved when, tempted by the deliciousness of local peaches, he ate what he thought were too many and yet stayed healthy. Titan Leeds advised readers in 1715:

By Apples, Nuts, Oyls, Eels [give] cold in head
By eating green fruit raw is hoarseness bred
The healthy may eat Cheese, and find it good,
But for the weak 'tis not such wholsome food.

Foods had their characteristic effects on bodily humors and people would do well to adjust their food to their own particular constitutions. Mustard seed, advised Daniel Leeds, was not good for choleric people. When the Hackensack Algonquian remarked upon colonists teaching Indians "debauchery and excess," Dankers understood it as an utterly logical explanation for the decline in Indian population in and around New York. Without attentiveness to moderation, no one could expect health.[10]

People had to adjust their choice of food to the change of seasons as well. In 1715 Titan Leeds wrote:

In Summers heat (when Choler hath Dominion)
Cool meats and moist are best in my opinion;
The Fall is like the Spring, but endeth colder,
With Wine and Spice the Winter may be bolder.

Such advice was a staple of almanac advice, appearing over and over. Just as the excess moisture and heat that unbalanced the humors or the toxins that corrupted them could move from the interior of people's bodies to the exterior through the skin via sweat or pus, so did the temperature and moisture of the surrounding air diffuse through the skin into the body's interior. For early modern Europeans, the skin was not a barrier separating one's self from the world, but a permeable borderland. A visitor to the ailing Mary Livingston reported to her father that she had recently "caught an Ague, by being a little too free with the Aire." With Mary's constitution already out of balance, she had worsened it further by moving from the protection of the house out into the heat of June. James Leigh, a carpenter, observed one February that his acquaintance William Dobbs had a fresh wound, and advised him to "binde a Hankerchiefe about it to keep it from ye Cold." Even the healthy had to moderate exertion, food, and drink to compensate for the threat from the extremes of the seasons.[11]

Healing too should be coordinated with the seasons. Everyone knew that winter, cold and dry, made the blood thin and scanty. Daniel and Titan Leeds regularly warned their readers not to allow themselves to be bled in January or February. By contrast, the heat and moisture of summer naturally engorged the body with blood. "Tis good in March to Bleed, to Bathe, to Purge," advised Titan Leeds in 1715, for this reduced the humors and prepared one's body for the inevitable excesses of summer. On the other hand, if by chance one had not lessened one's humors beforehand or sweated them down during the summer, September was the time to purge or bleed oneself back to balance.[12]

As when Dankers reported his conversation with the Hackensack Algonquian about the effects of immoderate living on Indians, colonists in New York did occasionally observe the health practices of the Indians around them. Charles Wooley described how the Algonquians coated their skin with fish oil, eagle fat, or raccoon grease to ward off sunburn and mosquitoes in summer; these substances were also a great "stopper of the Pores of their Bodies against the Winter's cold." He had a conversation with an Algonquian healer in which they compared their respective methods of healing the body through sweating. While colonists left the ailing in their beds and rubbed them with warm cloths, ailing Algonquians sat in sweat lodges and then plunged into a cold river. Wooley reported that the Algonquian healer argued for the advantages of washing off "the excrematitious remainder of the Sweat," but Wooley saw the dangers in closing up the pores of the skin precipitously. The Algonquians' practice, he noted, in particular "prov'd Epidemical in Smallpox, by hindering them [the pox] from coming out." Wooley understood smallpox to be a corrupter of humors, and that to survive one had to vent the corrupted humors through the skin. By closing off the natural egress, he thought, Algonquians doomed themselves to die.[13]

If Wooley was critical of Indians' healing methods in smallpox, on occasion colonists observed practices of which some of them approved. For example, several men remarked upon the ease with which Mohawk women apparently went through childbirth. Immediately after the birth, Mohawk women went to a nearby river ("Whether Winter or Summer," marveled one), bathed themselves and their newborn infant, and returned to their ordinary tasks. Indeed, reported the Reverend Johannes Megapolensis from Rensselaerwyck near Albany, "we sometimes try to persuade our wives to lay-in so, and that the way of lying-in in Holland is a mere fiddle-faddle." There are no reports, however, that colonial women were so persuaded, and nothing in these sorts of observations undermined colonists' convictions that they understood how the body worked and illness happened.[14]

For colonists in New Netherland and New York, then, like early modern Europeans elsewhere, their bodies were intimately connected with the surrounding natural world. Humors coursed through their bodies, sometimes scanty, sometimes to excess, sometimes sweet and wholesome, sometimes corrupted, dried, or stopped. The balance that was health varied from person to person, depending on one's sex and age, and shaped distinctive temperaments. But everybody's balance was changeable, affected by the air around him or her, the foods they ate, the liquids they drank, the effort they exerted. Balance was precarious and the possibility of slipping from health to illness ever present.

*I*f colonists' bodies were influenced by shifts in the seasons, seasonal shifts in weather were in turn influenced by more distant changes. In their almanacs, as in all colonial almanacs except those of early New England, Daniel and Titan Leeds offered predictions of the weather, for, Daniel asked rhetorically, "[W]ho can say that I do not as often judge right of the ruling

Humor in the great World, as the Doctors do of the ruling Humor in [MAN] the little World." Daniel and Titan judged the ruling humor by following the movements of the five known planets, Saturn, Venus, Mars, Mercury, and Jupiter. For January 1709, Daniel wrote

> Saturn in Northern Signs was found of Old,
> To cause our Winters to be long and Cold;
> And several years to come, he is placed so,
> Therefore he'll send us store of Frost and Snow.
> But for some Months still Mars in Cancer stays,
> And makes this Winter mild, as reason says.

Saturn, cold and dry by nature, exerted his force to exacerbate the winter overall, yet briefly Mars, warm and dry, would moderate Saturn's influence. The basis for Daniel Leeds's predictions was his knowledge of astrology.[15]

Astrology was based on the belief that the planets, as they moved in and out of positions of proximity to each other and in relationships to the twelve sections of the sky associated with the zodiac, systematically influenced the natural world. Astrology was developed out of the cosmology attributed to Ptolemy, a second-century librarian in Alexandria who compiled Greek and Roman learning. In the cosmology of Ptolemy, Earth was suspended at the center of the universe. Each of the seven heavenly bodies observed to have significant motion (the Sun, the Moon, and the five planets) was embedded in a transparent crystal sphere that rotated at a distinctive rate around Earth. These spheres nested one inside the other, rather like Russian nesting dolls, with Earth at the core. The fixed stars (fixed, that is, in relationship to each other) were embedded in the outermost crystalline sphere "like so many Golden Nails driven into the Top of some arched Roof." This sphere, too, revolved around Earth. The rotations of each of these spheres, each at its own pace, gave Earth a continuously changing pattern in the night sky. And the effects of these celestial bodies were, as a sermon printed in New York put it, "obvious to all men of ordinary Understanding."[16]

The Ptolemaic model matched ordinary observations of the heavens very nicely. When Daniel Leeds predicted an eclipse in 1699 as occurring "when the Moon is under the Earth," he was assuming a universe consonant with everyday experience, one in which he was "right side up," and the sun, moon, planets, and stars plainly arched overhead and then circled under the earth before arching overhead again. His readers, too, had considerable familiarity with the sky. They watched the sun mark the passage of days, the pattern of the stars, and the changes of the seasons. Not only did they use almanacs as a monthly calendar but as a daily timekeeper. Leeds, like other almanac makers, included the times of sun rise and set, of high tides, and of "southings" (passages across the meridian) of various familiar celestial bodies. These, he wrote, "may be Useful to planters, Water men, &c to know the Time of the Night." Rarely did colonists note what Indians knew about the sky, although an early colonist observed that Algonquian women planted when they saw the seasonal reappearance of a particular constellation. But colonists assumed they knew much more about celestial events than did Indians. When told that Indians couldn't predict eclipses, the Albany trader Robbert Sanders replied, "Ask my smallest girl when such things will happen 10, even 20 years from now. She will be able to tell you in less than an hour." In 1707 Daniel Leeds mocked the almanac of a Philadelphia rival Jacob Taylor because the Philadelphian listed the full moon as southing at the same time two nights in a row. "[A]lmost every Water-Man, and everyone that can but read," snorted Leeds, "knows the moon is ordinarily South a quarter of an hour later every night, as well at full as at any other time." While the origins of the Ptolemaic model of the heavens lay in classical learning, the everyday experiences of ordinary colonists continually affirmed it as true.[17]

Leeds presented his predictions of the weather as based on concrete phenomena that were accessible to anyone. Astrology, like any cultural practice, has changed historically, and in seventeenth century England a broad cultural effort to align knowledge more carefully with the observable natural world had an impact on astrology. During the turmoil of the English Civil War, astrological predictions had been deployed as political weapons, and as a consequence astrology's more extravagant claims had been discredited. Many of the arcane details and complicated decoding of previous centuries were stripped away, resulting in a simpler astrology that claimed to be closer to what actually happened in the skies. As well as trying to render astrology apolitical, these changes were also intended to divest astrology of any taint of magic. Daniel Leeds was born in England in 1652 and thus came of age during the height of this effort to strip astrology of any mysteriousness or occult power. Indeed, Leeds presented his skills as a straightforward matter of knowledge and experience. In 1699 he presented readers with a chart by which they too could figure out what weather was likely, given the knowledge of the planets' positions. Astrology, wrote his son thirty years later, "is nothing but a Bundle of Experience, which the industrious Observators have heaped up as a Portion and Legacy to After-Ages."[18]

To predict the effects of the planets on the weather, almanac makers like Daniel and Titan Leeds did not necessarily need very complex skills.

Colonial almanac makers usually didn't do much of the calculations of future celestial positions themselves, much less the numerous and careful astronomical observations that served as the basis for such calculations. Rather, they generally used ephemerides (tables of the future positions of the major celestial bodies) printed in Europe, as Daniel Leeds initially did, merely adjusting them to local time. But such ephemerides were not always available, given the vagaries of the circulation of print in the early colonies. By 1708 Leeds's copies had expired and he was forced to do his own calculations for the year, projecting 1708's positions based on celestial positions of 1707. Leeds used this opportunity, though, to promote himself, pointing to his own superior skills compared with his rival Taylor who, Leeds claimed, had "been lazily transcribing from a German Ephemeris, that still continues till 1712."[19]

Whatever Leeds's mathematical skills, the usefulness of his almanac as a timekeeper needed to be only approximately correct in order to suit the needs of his readers, for very few of them owned a clock against which his forecasts could be compared. On the other hand, egregious errors were plain to everyone. In 1697 John Clapp, a sea captain turned tavern owner in New York City, published an almanac in which the times of the year's eclipses were clearly wrong. That was the last almanac published by John Clapp. Colonists needed to be assured that their almanacs were reliable timekeepers and useful guides to daily work in healing and farming, if not necessarily precise.[20]

While the positions of the planets affected the weather and so human health indirectly, the position of the moon directly affected the human body. As the moon moved through different signs of the zodiac, it exerted influence over specific parts of the body. Almost all colonial almanacs had a woodcut of the "Man of Signs" ("The Anatomy"), which displayed the connections between each sign of the zodiac and the part of the body over which it "ruled" (illustration 4). In the monthly calendar, there was a column for the zodiac position of the moon for each day of the month. Healing was most efficacious when coordinated with the moon. For example, it was "held to be extream perilous for the Moon to be in that sign which ruleth the Member where the Vein is opened," *The Husband-man's Guide* reminded readers. So if the moon was in Gemini on a given day, one wouldn't draw blood from an arm, for arms were ruled by that zodiacal sign, but would choose another site on the body. By contrast, applying an ointment or plaster was best done when the moon was in the ruling sign for the body part affected. By using the "Man of Signs" and the table of the moon's positions, almanac readers

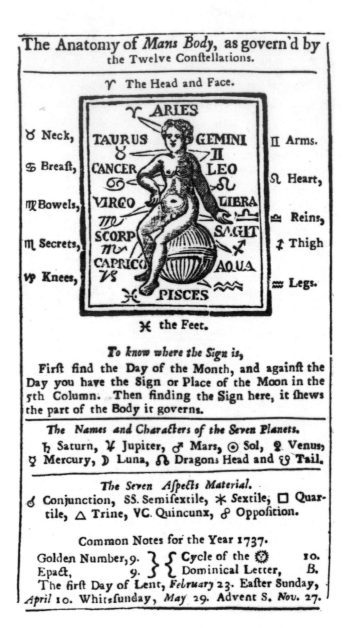

The Anatomy of *Mans Body*, as govern'd by the Twelve Conſtellations.

♈ The Head and Face.

ARIES

♉ Neck,   TAURUS   GEMINI   ♊ Arms.

♋ Breaſt,   CANCER   LEO   ♌ Heart,

♍ Bowels,   VIRGO   LIBRA   ♎ Reins,

♏ Secrets,   SCORP   SAGIT   ♐ Thigh

♑ Knees,   CAPRICO   AQUA   ♒ Legs.

PISCES

♓ the Feet.

To know where the Sign is,

Firſt find the Day of the Month, and againſt the Day you have the Sign or Place of the Moon in the 5th Column. Then finding the Sign here, it ſhews the part of the Body it governs.

The Names and Characters of the Seven Planets.

♄ Saturn, ♃ Jupiter, ♂ Mars, ☉ Sol, ♀ Venus, ☿ Mercury, ☽ Luna, ☊ Dragons Head and ☋ Tail.

The Seven Aspects Material.

☌ Conjunction, SS. Semiſextile, ✳ Sextile, ☐ Quartile, △ Trine, VC. Quincunx, ☍ Oppoſition.

Common Notes for the Year 1737.

Golden Number, 9. } { Cycle of the ☉   10.
Epact,   9. } { Dominical Letter,   B.
The firſt Day of Lent, *February* 23. Eaſter Sunday, *April* 10. Whitſunday, *May* 29. Advent S. *Nov.* 27.

ILLUSTRATION 4. "Man of Signs" ("The Anatomy"), in Titan Leeds, *The American Almanack for the Year of Christian Account 1737* (Philadelphia: Andrew Bradford, n.d.). Almost ubiquitous in colonial almanacs, the seated figure is surrounded by the signs of the zodiac, each of which was thought to govern a particular part of the body. *(Reproduced with permission of the American Antiquarian Society.)*

could recognize the likely influence of the moon on their own bodily state and act accordingly.[21]

Other living things were similarly subject to celestial influence, and so astrology was useful in agriculture too. Commonly circulated print in early New York advised readers to time their farming to take advantage of the influence of the heavens on their livestock and their crops. Geld sheep and cattle when the moon is in Aries, Sagittarius, or Capricorn, shear sheep when the moon is waxing in Taurus, Virgo, or Libra, and kill swine at or near the new moon, advised *The Husband-man's Guide*, so the "flesh will prove better in boyling." Specific plants were influenced by particular planets and so had particular qualities. Daniel and Titan Leeds periodically gave their almanac readers lists of herbs commonly used in medicinal recipes. "Tis the Opinion of Astrological Physitians as well my belief," wrote Daniel Leeds in 1697, "That all Herbs are stronger, and of greater virtue, if gathered when the Planet that governs them is well fortified." Consequently, his almanacs included each herb's ruling planet and the dates that year when they would best be gathered. Crops, too, did best if coordinated with the skies. Sow peas, beans, and parsnips with "the Moon decreasing," advised *The Husband-man's Guide*. That at least some readers in New York followed this sort of advice is indicated by a reader's note in the margin of a 1705 almanac: "The first full Moon that's in July/Sow turnip seed [Ye't moist?] or Dry." Similarly, Philip Livingston wrote his father in January 1724 that he hoped the new moon would bring him "good winter wheather," for timber felled in January lasted longest and timber felled near the new moon was best of all.[22]

While soil was not imagined as directly subject to astrological effects, still it shared some qualities with living plants and animals. As with the pores of the skin, cold closed the earth and warmth opened it. In June, advised Daniel Leeds, lay hay, straw, or weeds around the roots of orchard trees to keep the earth "moist and open." However, in their attentiveness to local soil, colonists wavered between regarding conditions in New York as no different from anywhere else and recognizing the specificity of their location. As soon as "the Earth is open," wrote Daniel Leeds on his February page for 1695, readers could go ahead and sow hardy seeds and pull up tree stumps. While this may have been standard advice in England's milder climate, nowhere in New York could one sow seeds in February. Daniel Denton claimed that Long Island soil was "very natural for all sorts of English Grain" as well as other crops. But Daniel Leeds remarked that English wheat grew well initially in the colony "but grows lean when sown on fallow Ground, the Earth being but as a step-mother thereunto, yet Indian Corn, being Natural, seems never to wear out [illegible] nor the Land whereon tis yearly planted." However much colonists transplanted to New York saw the same effects of celestial bodies on the natural world around them that they would have had they stayed in Europe, some aspects of the local natural world seemed particular to this place.[23]

*E*ven though by around 1700 some people had worked their way into what constituted a political and economic elite within the colony, they did not on that account withdraw from common beliefs about and practices in the natural world. Neither Robert Livingston nor Alida Schuyler Livingston had much education, although both were literate and numerate, and though they developed significant wealth, their letters exhibit no departure from the natural world Daniel and Titan Leeds depicted. Lewis Morris, born in New York City in 1671 to parents who had recently moved there from Barbados, was lucky enough to inherit considerable fertile acreage when he was twenty. He, too, had had only a modest formal education, although he would go on to accumulate a library with three thousand volumes, and self-consciously cultivated the life of a gentleman distinct from what he saw as the boorish colonists around him. For instance, he mocked the concerns of most of those elected to the General Assembly, the colony's legislative body.

> Their great debates, of piggs and fowl
> What bigness Stallions ought to Strou'l?
> If hogs should run at large?
> Concerning foxes slaying Pullen,
> Which made the good wives very Sullen;
> And cattle that were stray'd.
> Such things took up most of their time,
> And tis not worth a verse rhyme
> To tell what each man said.

Nonetheless, he too worked to develop his estates, and his letters are full of the identical concerns, as when he cautioned a son, embarking on his own agricultural ventures, to attend to whether the soil was "mellow" or "foule." He cautioned another son recovering from an illness "not to be Exposed to the night Air nor to the wet," for fear of setting back his recovery. Whatever Morris's pretensions, his understandings of the body and the natural world were like everyone else's. Like the universality of the experience of one's own health and illness, experience with animals and plants, and the difficulties of making them thrive, was virtually everyone's.[24]

As was true in most colonies, very few upper-class or extensively educated Europeans ever appeared in New York. An occasional governor, like Lord Cornbury, governor between 1702 and 1709 and cousin to Queen Anne, moved in truly exalted circles in London. Ministers dispatched to serve colonial churches generally were among the few who came with a university education, men like Megapolensis who arrived in 1642 with a degree from the University of Cologne. Only an occasional other colonist emigrated with such an education. One such colonist was Jacques Cortelyou, with a degree from the University of Utrecht. Cortelyou was hired in 1652 to serve as a tutor to the children of a wealthy merchant and stayed to farm on Long Island. Jaspar Dankers visited him in 1680 and described him as knowledgeable in mathematics and Latin, as well as versed in (regrettably, in Dankers's opinion) the philosophy of Descartes. But Cortelyou was exceptional among those colonists born in Europe. Most came with modest educations, if any. In New Netherland and New York itself, scattered schoolmasters and occasional Latin masters offered learning beyond the literacy taught at home and in dame schools (run by women in their homes). But few others had the wealth and the leisure even for that. Absent a substantial number of people of wealth and leisure, the distribution of learning in New Netherland and New York roughly followed the social distribution of colonists: by and large, the middling and lower sorts with middling educations or less.[25]

But erudition was not much missed, as few among the colonists looked to formal erudition or extensive expertise in others in dealing with the natural world. Even in medicine, what expertise there was was largely an elaboration of what everyone knew for themselves. In Europe, there was an erudite medicine institutionalized in a tripartite social structure of university-trained physicians who did the intellectual work of assessing illnesses and prescribing for them, apprentice-trained surgeons who did physical work like bleeding and bone-setting, and apothecaries who compounded medicines. But even in Europe, few outside the wealthy could routinely afford such services. Most people treated themselves, relied on kin and neighbors, and paid for assistance from healers of variable training and experience only when warranted. In the colonies, Johannes Kerfbyle, who immigrated to New York with a medical degree from the University of Leyden in 1685, was a rare example of a university-trained doctor. In New Netherland and New York, university-trained physicians were few and far between.[26]

Most of those with healing skills in the colony who appear in the seventeenth-century and early eighteenth-century records were described as midwives or surgeons. These latter were sometimes paid by towns to treat the indigent ill, and colonists called upon them for physical procedures, as when Harmen van den Bogaert incised a fellow traveler's swollen leg or when Jacob de Hinse incised Maria van Rensselaer's painful hip, releasing a piece of matter "as thick as one's thumb and four inches long." As time passed, colonists were less and less likely to distinguish between physicians and surgeons. In 1698 Albany-born Jacobus Provoost was admitted to the privileges of the town, with his occupation listed as "Physitian and Chirugeon," although he could have done no more than apprentice locally. Lewis Morris mocked a local doctor whom he described as

> A tooth drawer, there [in Europe] not worth a groat
> Who run on tick or Stole a coat
> Proves here a learn'd Phisician.[27]

But whatever Morris's sense of the erudition and gentlemanly appearance proper to a physician, in the colonies "doctor" was a term given to any man who engaged regularly in healing work, no matter what his education. Morris himself trusted the advice of a man trained as an apothecary, not a university-trained physician. One of the men accused of participating in the 1712 slave uprising, in fact, was referred to as "Peter the doctor," indicating that healing work was part of the repertoire of skills of a wide range of men. Similarly, investigations of unexpected deaths often included ordinary men, for death was so familiar an experience that reasonable judgments were assumed to be within their realm. Thus seven jurymen in East-Hampton in 1670, finding a mark on the face of a drowned boy, determined that he had likely hit his head while diving into the water, and twelve jurymen in Albany in 1700, finding no mark or bruise on a man found dead one morning in his canoe, pronounced the death as natural. Still, colonists were aware that erudition carried weight elsewhere. When Governor Henry Sloughter died suddenly, the investigation was performed, the Governor's Council assured London, "by the Phisetians and Churugeons of the place." Clearly, this was no time for the judgment of common jurymen. But within New Netherland and early New York generally, what expertise there was lay in skills that were sufficiently like the skills of the everyday that ordinary people could often enough approximate them.[28]

Nonetheless, someone who was perceived to be especially knowledgeable and skilled could be sought out for help. Jacques Cortelyou, for example, was apparently personally respected and was thus called upon repeatedly by neighbors. When Dankers first visited him, Cortelyou was absorbed in caring for his own sons, ill with smallpox, but on a return visit

Dankers saw him helping a neighbor whose horse had "the staggers." Care of such ailing horses entailed bleeding them and administering warm drinks, much like caring for people (although Dankers also reported that sometimes New Yorkers made "a puncture in the [horse's] forehead, from which a large quantity of matter runs out," something presumably not applied to humans). When a group of families decided to establish the town of Schenectady on the Mohawk River, they trusted Cortelyou to lay out their plots of land fairly and judiciously, and his surveys indeed held.[29]

Mere erudition, however, was not enough. By 1700 Johannes Kerfbyle, the Leyden-trained physician, had sufficient social standing to serve on the Governor's Council, had been the lone man with a formal MD to serve on the investigation into Sloughter's death, and served as an elder in his church. But he apparently had a net worth of only a modest twenty-five pounds, an indication that only a similarly modest number of New Yorkers had found his erudition worth purchasing. Sharing a commonsense understanding of the sources of illness and the routes to health, just as they knew how to manage their crops and animals, ordinary colonists turned to the unusually well-educated only when they trusted those men's good sense and experience. They admired men like Cortelyou not because he possessed an arcane body of knowledge but because he was a man whose reliability was based on knowledge that was simply an extension of their own.[30]

Indeed, a counterpoint to acceptance of learned authority was the occasionally expressed conviction that ordinary people were in fact *more* knowledgeable than the book learned. "The meanest things are often found to cure the [seeming] most incurable Maladies, when the richer things have failed," wrote Daniel Leeds at one point. "[A]nd," he added, "many silly good Old Women have out-done the most profound Physitians in effecting cures." Among colonists, the economy of knowledge was one in which people were trusted based on what they could do, and on perceptions of their character. Formal credentials were of relatively little moment.[31]

*L*eeds's reference to "Old Women" raises the question of women's participation in this economy of knowledge. Historians have often pointed out that in the early modern period European women were explicitly relegated in both Christian theology and political philosophy to subservient positions within households dominated by men. Simultaneously, European writers and artists symbolically associated nature and women, both in the sense that "nature" was often personified as a woman and that women's childbearing was metaphorically assimilated to natural processes like bearing fruit.

Women's association with nature reinforced their social subordination, as men were charged with mastery over both households and the natural world. Colonists in seventeenth century New Netherland and New York rarely speculated about nature in general, and so it is hard to recover the degree to which women in the colony were regarded as more akin to the natural world than were men. It may be that this was so ordinary an assumption that no one bothered to express it. Daniel Leeds has left us a poem about reproduction, which is as follows:

> The first six days, like milk, the fruitful seed,
> Injected in the Womb, remaineth still;
> And other nine, of milk, red Blood does breed:
> Twelve days, turns blood to flesh by nature's Skill;
> Twice nine, firm parts, the rest, ripe Birth doth make
> Such Shape and form, in time, man doth partake.

Here, Leeds's images are those of a fermenting, changeable natural world, in which nature itself seems to have no gender. But since few people in the colony speculated in print about "nature" as a concept, it is difficult to know the degree to which these colonists feminized the natural world.[32]

Whatever the clarity of this complex of beliefs about women as lesser and subservient in the early modern Atlantic world generally, in this particular colony women were quite assertive. Dutch law gave women, particularly married women, more autonomy and more rights over property than did English law. Long after the English took the colony in 1664, Dutch legal practices persisted. As elsewhere in the early modern world, widowhood or the absence of husbands away soldiering, sailing, or trading meant that women, in fact if not in theory, often ran family enterprises for periods of time. In this colony in particular, the strong orientation toward commerce gave women opportunities for an independent economic existence, if often enough a precarious one, as traders, shopkeepers, or tavern keepers. New Yorkers' common Protestantism, while advocating decorous and subservient behavior, nevertheless asserted an equality of spiritual life and fostered literacy among all its congregants. By the latter half of the seventeenth century, women were notably active in New York's churches. Overall, court records and personal papers show that women in New Netherland and early New York were as active and engaged as they were anywhere in the early modern Atlantic world.[33]

Still, there were some areas related to mastery of the natural world in which women had no formal role, which were areas that required mathematics: navigation, surveying, almanac making. But note that it was a woman who was disputing her property line in the incident in the previous chapter in which the surveyors could not agree, and that Robbert Sanders's smallest girl knew how to calculate an eclipse. So the absence of women in these public roles hardly means such things were a mystery to them. Healing, on the other hand, was a household task shared by men and women in similar ways, as when Cortelyou nursed his sons. Nevertheless, only men who engaged in paid healing work were addressed as "doctor," with women's paid healing work largely limited to midwifery, although this was exclusively theirs. In agriculture, men and women's tasks were largely distinct, with men tending field crops and livestock and women working in gardens, processing food, and making cloth. Both would have drawn similar sorts of information about coordinating their tasks with the heavens. Overall, then, while some tasks were not practiced by everyone, understandings of how the natural world worked and what one should do accordingly were broadly shared by men and women alike.[34]

For all of them, the truth of their understandings was constantly confirmed by everyday experiences. People's bodies felt akin to the natural world around them. The sense of the body as permeated by fluids and surrounded by permeable skin was confirmed by runny noses and seeping sores. In a period in which people seldom washed, bodies stank like the rotting, fermenting, and musty natural world around them. Close contact with animals confirmed their similarity to human bodies.

Hard work with plants and soil required attentiveness to their immediate and specific qualities. Everyday observation confirmed the movement of the stars, planets, sun, and moon, all rising in the east, arching overhead, and setting again, all circling around an earth that stayed plain and still at the center of them all. Work confirmed the necessity of personal experience amid the vagaries of chance: weather, illness, the loss of an animal could shift one's fortunes in an instant. Print offered hope for adding to one's repertoire of skills in agriculture, trade, and healing, but experience showed that nothing worked all the time. As Daniel Leeds wrote,

> The Doctor cannot always help the Ill.
> The Sickness sometimes is beyond his Skill.

While Leeds recognized that some of his calculations for planetary motions were only approximations, he wrote that they were "exact enough for common Use," and indeed they were. Any dissonance between print advice and its applicability to actual practice disappeared because of colonists' modest expectations for the "truth" of knowledge about the natural world.[35]

CHAPTER 3

Anomalies

God rules this world by Stars and Angels, so
As Angels Countermand what Stars can do
In general judgments, Angels power's known;
Our single fate he rules by Stars alone;
For where's the man, who with success has Strove
Against the power of Mars, Sol and Jove?
　　　Daniel Leeds, The American Almanack For . . . 1708

On a winter's evening in 1657, Sam-
uel Parsons dropped by the house of Arthur and Elizabeth Howells in the
Long Island town of East-Hampton. Mrs. Howells was sitting by the fire,
and when Parsons inquired how she was, "she said she thought she had
gotten some could [cold]." Feverish, she put on lighter clothes and bound
a cloth around her aching head. Her husband, just then coming in with a
neighbor, prevailed upon her to go to bed. Mrs. Howells nursed her infant
daughter, but after Parsons took the child from her, she suddenly turned her
palms upwards and shrieked, "[A] witch a witch: now you are come to torter
me because I spoke 2 or 3 words against you!" Parsons, "affrighted at her
being taken sudenly in so strange a maner," said to the neighbor, "[T]he
lord be mercifull to her." But Mrs. Howells was taken with a mortal illness.
Despite the care of her husband, her mother, her slave woman, and several
neighbors, both women and men, she succumbed within days, leaving her
family bereaved and a local woman to stand accused of witchcraft.[1]

Colonists in New Netherland and New York lived in a natural world
suffused with spiritual forces. "God rules this world," wrote Daniel Leeds,
and indeed colonists believed that the regularities and relationships they
experienced were his direct creation. So, too, were unexpected events
and things, the anomalies that broke through the tissue of the everyday.
Interruptions in ordinary patterns, such as eclipses and epidemics, were
interpreted as instances of God's Providence, messages of his judgment
that called them to repentance and reform. The Devil was also an active

presence in the material world, and on occasion colonists would accuse one of their own of witchcraft, of being in the Devil's service and not in God's. Colonists regularly described the spiritual practices of the Algonquians and the Iroquois as Devil-worship, confirming their place outside a Christian society. They noted, too, the magical practices of Africans among them, and colonists consistently defined their slaves as heathens, in contrast to their own identity as Christians. On the other hand, occasional instances of oddness seemed neither part of God's Providence nor the Devil's work, but were simply strange, so-called "wonders." Yet even these seemed to affirm the widely shared understanding that the natural world was pervaded by spiritual forces.

Because New Netherland was rather poor and attracted relatively few settlers, the development of formal religious structures in the colony was slow and uneven. The first Dutch Reformed minister arrived in New Amsterdam in 1628, and gradually other congregations were established in the areas around Manhattan and up the Hudson in Albany, Schenectady, and Kingston. As New Englanders migrated to Long Island beginning in the 1640s, they brought with them their own version of Calvinist worship. Non-Calvinist Protestants, like Quakers and Lutherans, established congregations beginning in the 1650s, as did a small group of Jews. When the English took the colony in 1664, they brought the Church of England. Huguenots, who settled on Staten Island and in Westchester County after their expulsion from France in 1685, brought their Reformed Church, and Presbyterians from Scotland and Ireland settled in increasing numbers as the seventeenth century turned into the eighteenth. Amidst all this diversity, first the Dutch West India Company and then the English government, both anxious to promote immigration from without and civil peace within, insisted upon a policy of toleration.[2]

Toleration was difficult to enforce, and indeed, like commercial competition, was one of the chronic sources of contention in the colony. The Reverend Megapolensis reported that in 1657, when two young Quaker women "began to quake and go into a frenzy," crying that the day of judgment was at hand, they were promptly jailed for disturbing the public peace. The depth of antagonism that Daniel Leeds, who had migrated as a Quaker but subsequently repudiated Quakerism, held towards his former coreligionists is indicated by the title of a pamphlet he published in 1701: "News of a Strumpet Co-Habiting in the Wilderness or, A brief Abstract of the Spiritual and Carnal Whoredoms and Adulteries of the Quakers in America." In 1658 Jeremias van Rensselaer's brother Richard wrote him from Amsterdam that their brother Nicholas had abandoned his business, claiming that "he has the spirit of truth, that in his dreams he sees many visions." Richard confided, "We fear that he is half crazy," but Nicholas reported one of his visions, that Charles II would soon regain the throne of England, to Charles himself. When Charles did in fact regain his throne in 1660, Nicholas showed up in Albany with a royal appointment as minister to the local Dutch Reformed Church. The congregants were not pleased.[3]

Despite official toleration, some governors felt that the Church of England was entitled to preeminence in the colony, and other groups subject to some control. For example, Governor Cornbury, who served from 1701 to 1708, strengthened the finances of Trinity Church, the Anglican congregation in New York City, and appointed Anglican ministers to non-Anglican churches in some of the surrounding counties. He also insisted that he was authorized to determine who was and was not a legitimate minister in other denominations. His actions infuriated New Yorkers of other persuasions. In a different vein, in 1713 the rector and vestrymen of Trinity Church felt compelled to protest "some Busey mockers & scoffers of Religion, who Ridicule both Sacred things & Orders by their profane Lampoons." Religious allegiances, even if almost uniformly Protestant and Christian, nevertheless were experienced by colonists in New Netherland and New York as a source of difference and a subject for disputes.[4]

There were some efforts to Christianize Africans and Indians in the colony. In the fluidity of the early decades of slavery, some slaves were baptized and married in the Dutch Reformed Church, but this became increasingly uncommon after the 1660s. Colonists worried that while enslaving heathens was congruent with Christian virtue, keeping a fellow Christian in slavery was not. Some slaves came into the colony already Christians, for in the Atlantic world people sold into slavery had diverse origins and experiences. Two of the slaves accused in the 1712 plot, for example, were "Spanish Indians" who were "most Suspected of having most understanding to carry on a plot" because they were Christians. These men had recently been captured from a Spanish ship by a New York privateer, and indeed maintained that they were neither slaves nor guilty of participating in the rebellion. In 1703 Elias Neau, a Huguenot, opened a school in New York City under the auspices of the Anglican Society for the Propagation of the Gospel through which he offered Christian teachings to slaves. In the wake of the 1712 slave rebellion, for which Neau was popularly blamed (even though only two of the accused had ever attended his school, and only one of them "was baptized and he was Condemned on Slender

Evidence"), public pressure closed his school, although it quietly reopened later. Thereafter small but steady numbers of slaves, in New York City at least, were baptized in the Anglican Church. Similarly, ministers occasionally tried to Christianize local Indians. A minister in Albany managed to baptize a few Mohawks in the 1640s, and forty years later another minister had somewhat more success. By the early eighteenth century perhaps a hundred Mohawks attended church regularly. But these efforts to extend Christianity to Africans and Indians were, by and large, individual and unusual efforts.[5]

Clergy in the colony primarily concentrated on their own congregants. As described in chapter 1, in the eyes of ministers, governors, and other men of authority, too many colonists were deficient in Christian decorum. Some apparently openly scoffed at religion, as when a Queens justice of the peace, Jonathan Whitehead, was accused in 1703 of saying that "Religion was onely an Invention of cunning men to gett thaire living by." Despite the steady development of a landscape sacralized with churches and despite ever-rising levels of churchgoing, most ministers believed that tending to the colonists required all their efforts.[6]

The diversity of churches and the difficulties of enforcing public decorum did not mean that religious allegiances were necessarily shallow. A case of political conflict in 1689–1691 illustrates the depths of religious convictions among many colonists. In the turmoil surrounding the 1689 assumption of the English throne by William and Mary, political control in New York was seized by a wealthy merchant and militia captain, Jacob Leisler. His governance elicited both strong support and vehement opposition within the colony. While this conflict partially stemmed from long-standing political and economic disputes over taxation, inheritances, and municipal privileges, Leisler's efforts on behalf of his particular version of zealous Calvinism was no small factor in the consequent discord. While the issues did not all turn upon religion, nevertheless both sides understood it in passionately religious terms. An opponent of Leisler described his rule as "the production of a Monster begat by an Incubus on a Scotch Witch, who had kindled his malice against Truth from the flames he put to the holy Bible." Leisler's supporters were equally sure that their opponents were godless sinners. After the execution of Leisler and his chief lieutenant by a governor freshly dispatched from London, a supporter safely in Boston described the executioners as "abominable filthy wretches . . . who have murthered, and persecuted two true Protestants to death, whose blood is crying under the alter." By the beginning of the eighteenth century, then,

New Yorkers were religiously diverse but not on that account either mild or perfunctory about their religion.[7]

*D*espite all this diversity and dispute, the Protestantism of New Netherlanders and New Yorkers was sufficiently alike that they shared a broad set of beliefs about how God acted in the natural world. Whatever the peculiarities of the local natural world, colonists never doubted that both the usual and the anomalous were under the constant supervision of God. "Providence" was the term early modern Europeans used to describe God's interventions in the world, interventions that conveyed his judgments on human affairs. "[I]t rains every day and the ground is so wet we can neither harrow nor plow," Alida Livingston wrote her husband in 1720. "I hope the Lord will turn his wrath from us and give us decent weather." God's Providence could indicate his favorable judgment as well. "[A]s he was eminently qualified for Public Imployments," said the Reverend William Vesey of a recently deceased governor, "so the Providence of God soon introduced him on the stage of Action." But even troubles could be a mark of God's favor, for Providence was a flexible and capacious idea. "[T]he Lord chastises whom he loves and punishes every son whom He adopts," wrote Maria van Rensselaer. "If this had not been my joy and strength, I should long ago have perished in my sorrow." Flood or fortune, all was an outcome of God's will.[8]

Illness, while having the natural causes described in chapter 2, had as its first cause God's judgment on a sinful people. Despite efforts to prevent illness by adjusting one's diet to the seasons and purging appropriately, and despite their repertoire of healing methods, colonists understood illness generally as part of the human inheritance from Adam and Eve's fall, a reminder of people's consequent inherent sinfulness. And while it behooved people to bestir themselves on their own behalf, ultimately relief from any illness came only with the help of God. The "good Lord grant you patience and strength to bear it and send you relief from it, for vain is the help of man," wrote Dr. John Nicoll of New York City to Robert Livingston in Albany in 1725, even as he enclosed medicine and advice. Titan Leeds rhymed in 1715:

> As Agues shake the joynts of feeble Men,
> So Discontent doth vex some now and then.
> But Scripture saith, with Patience bear your cross,
> And think it gain to undergo some loss.

Despite Jaspar Dankers's disapproval of many of the New Yorkers he encountered, he admired Maria van Rensselaer because of her acceptance of long years of suffering with a painful hip, her illness merely one of the chastisements God had sent her. As Philip Livingston wrote to his father, commenting on the unexpected deaths of three people hale and hearty but a few days before, "God grant us grace to be all ways prepared for that great Change." Illness was part of everyone's lot, a message that was not directed at those who particularly sinned, but one that came to the devout and the wayward alike.[9]

The messages of other events were less clear. Daniel Leeds, ever alert to the ill effects of Quakers, pointed out that an epidemic of blast in local orchards "bears date from the Time that the fiery Zeal of the QUAKERS broke out in a flame one against another." For him, the infected fruit trees bore witness to a diseased spiritual state, a natural world gone awry when the human world erred. In 1689 Leisler wrote to London of a fire kindled in New York City "suspected to be done by one Papist." Thankfully, a slave discovered the fire in time and residents were able to put it out before much harm was done. While their actions were the proximate cause of averting disaster, Leisler concluded that "the people wer trew God's mercy miraculously saved of that hellish design." Quakers clearly would dispute Leeds's reading of the orchard disease, and Leisler's political opponents would hardly have acknowledged his worthiness to interpret Providence correctly. While none doubted that God acted in both the natural world and in human affairs, the message that should be read in particular events was not altogether clear.[10]

But whether clear or not, repentance and reform was always the appropriate response. Anna van Rensselaer in Holland wrote her son Jeremias in Rensselaerwyck in 1659 that she was sorry that he had had bad luck with both merchandise and cattle. But he should consider, she admonished him, whether "you are not serving God as you should." Natural disasters in particular were anomalies that were instances of God's Providence and warned that human affairs were particularly disordered and called upon people in general to repent and reform. In 1695 a storm snapped off the mast and tore off the rudder of a ship carrying Robert Livingston to Europe. The crew and passengers drifted for weeks in the vast Atlantic. "We must admit we are an exceedingly godless lot of people, sinning daily beyond measure," Livingston wrote in his journal. Eventually they drifted onto the coast of Portugal, and Livingston wrote, "I thank God in prayer for His mercy and miraculous delivery." Natural disasters were occasions for colony-wide responses. In 1662 Peter Stuyvesant declared a day of fasting and prayer following one such series: epidemic disease, then heavy rains followed by drought that ruined the harvest. So too disasters elsewhere, as when an earthquake hit the island of Jamaica in 1692, were occasions to contemplate the consequences of sin and religious negligence.[11]

Little about the meaning of natural disasters was peculiar to New York, for the colonists' God operated everywhere the same. One of the few exceptions was the presence of rattlesnakes that so unnerved colonists. A truism in the medical advice that circulated popularly was that God providentially left cures for local ailments in the local natural world. Rattlesnake bite was one of the few instances in which colonists in New Netherland and New York sought out Indian knowledge, for while images of poisonous snakes were found in the Bible and in classical literature, colonists had had no experience with them before crossing the Atlantic. "[T]here grows spontaneously in the country, the true snake root," Adriaen van der Donck assured readers, "which is very highly esteemed by the Indians as an unfailing cure." But colonists found nothing that worked reliably. Charles Wooley, in the colony in the late 1670s, claimed that "all-wise Providence which hath furnish'd every Climate with antidotes proper for their distempers and annoyances, has afforded great plenty of Penny-royal" whose scent allegedly killed rattlesnakes. But he said nothing of a providential local plant to heal the bitten, instead describing Indians as sucking out the poison. In his 1695 advice to readers, Daniel Leeds stuck to the familiar European herb rue, as it was an "Enemy to Poyson" generally. Upon occasion, then, despite their conviction that their Christian beliefs were universally applicable, the local natural world sometimes defied their expectations.[12]

*O*f all the visible acts of Providence, celestial events were held to be the most pregnant with meaning. Comets, in particular, were unequivocally messages of God's wrath and harbingers of further disaster on the earth below. Jeremias van Rensselaer wrote his brother in 1665 that, for the moment, things were going well. "However," he added, "God Almighty has three several times showed us comets." What they will bring "we know not. We . . . hope that He for His Son's sake [illegible] Jesus Christ, will be merciful to us." Similarly, in 1681 Albany officials sent a letter to the acting governor about "ye Dreadfull Comett Starr" that had first appeared in the sky several weeks before. Like Van Rensselaer, they described its location, color, and the configuration of its tail because such attributes were supposed to be of help in decoding its message. Although the acting governor admitted that he couldn't specifically interpret the attributes, he wrote to the Albany officials, agreeing that "they certainly threaten God's Vengeance and Judgements and are as premonitors to us. Soe I doubt not," he continued, that each of them would do "your performance of your duty by prayer, etc. as becomes good Christians Especially at this time." More clearly than any other natural event, comets called for repentance and submission to whatever judgments would follow.[13]

Epidemic illness, unlike everyday illnesses, were anomalies akin to comets, for they too were believed to come at moments of collective sinfulness. From the Black Death onward, epidemics loomed large in the European imagination, dreaded events that erupted in communities and decimated them. A Dutch soldier wrote Governor Stuyvesant from Esopus in the Hudson Valley in 1664 of an epidemic that "increases daily." "It is a strange disease," the soldier wrote, "but the Almighty's will be done." By the late seventeenth century, smallpox became the most feared of these diseases, sometimes referred to as "the plague," though clearly distinguished from the Black Death plague. While prayers and sometimes communal fasting were believed to be appropriate responses to epidemics, settlers did not on that account neglect other sorts of actions. However much epidemics were a chastisement, their secondary causes were of the material world. Some were suspected to be caused by a "miasma," a putrefaction of the air that seeped through the skin; others clearly moved somehow from person to person. Consequently, colonists implemented practices learned through experiences with epidemics in Europe, such as clearing towns of "filth" and quarantining the ill. In the wake of an epidemic, the Albany Common Council in 1686 issued regulations on disposing of wastes in order to try to keep well water clean, and when smallpox broke out in the port of New York in 1702, the Albany councilmen, "sensible of ye daily visitations of Almighty God," forbid all movement between the two towns. But if epidemics called for material actions like cleanliness and quarantine, nonetheless they were like comets in that the call to repentance and reform was the primary message.[14]

As with Daniel Denton's comment in 1670 that "there is now but few" Indians on Long Island, colonists were intermittently aware that the number of Indians among and near them was declining. Everywhere in the Americas, natives of the hemisphere suffered catastrophically when newcomers from Europe and Africa arrived and brought contagious diseases, and the Algonquians and Iroquois in and near New York were no different. Isolated from the Old World where epidemic diseases flourished because of proximity to domestic animals, the density of human populations, and the constant movement of traders and migrants, natives of the New World were particularly vulnerable. Many of the newcomers to the Americas had survived bouts of infectious disease already, some of which, like smallpox, conferred immunity on survivors.

Such survivors, therefore, were able to nurse the sick and carry on the work of their communities when epidemics recurred. But in the first waves of epidemics, Indians got sick all at once, with no one well enough to cook or bring water or bathe the ailing, and the consequent death toll was enormous. Epidemics then recurred periodically. Indians also died during warfare or bouts of drunkenness, or they simply migrated away from colonial settlements rather than endure colonial encroachment. "As to the Indians," wrote the minister George Muirson of Rye in Westchester County in 1708, "they are a decaying people." Where there had been several hundred Indian families in his parish when he arrived some years before, he now saw less than twenty.[15]

Colonists in New York sometimes saw the slow decline of Indians on Long Island, Manhattan, and the lower Hudson as the outcome of Providence. "[I]t is to be admired," wrote Denton, "how strangely they have decreast by the Hand of God, since the English first setling of those parts." But others well recognized their own role. "It is to be feared," thought Adriaen van der Donck, that because of how colonists dealt with Indians, "on the Last Day they will stand up against us." Mostly, however, colonists in New York understood epidemics among Indians as little different from their own. When the Iroquois suffered from smallpox in 1716–1717, for example, Governor Robert Hunter offered his condolences, but also the advice that, since the smallpox was no doubt God's judgment upon them, the Iroquois would do well to repent of their sins.[16]

The Iroquois declined to take comfort from Hunter's advice because, as the Iroquois Dekanissore pointed out to Hunter, they understood smallpox differently. They believed that they suffered because of someone's malevolence and thus they would be better served if that person would desist. The Iroquois understood diseases to arise from an imbalance in the spiritual forces of the world. Such an imbalance could arise when someone failed to act correctly (as in violating a taboo or omitting an obligation), had an unfulfilled wish, or was the object of malevolence, whether randomly or for cause. Even illnesses to which Europeans would have ascribed a simple mechanical cause, such as a broken bone or a wound, involved issues of luck, of spiritual force for the Iroquois. Central to Iroquois practice were healing performances, ceremonies that might incorporate sweating or healing plants but whose larger intent was to restore balance to forces with both spiritual and corporeal manifestations.[17]

The Iroquois, all of whom lived west of the colony, seemed to have tried to adapt to the onslaught of epidemic diseases themselves, just as Europeans had in the wake of the Black Death. Their wars to the west and south all through the seventeenth century were often adoption wars, as they sought to replace their lost population by incorporating captives into their families. During the 1716 outbreak of smallpox, an English missionary complained that several Mohawk converts reverted to "their Old Heathenish practices" of ceremonial feasts and acting out wishes in dreams. The Pennsylvania botanist John Bartram, staying in an Onandaga village in the 1740s, described watching a man wearing "a clumsy vizard of wood colour'd black, with a nose four or five inches long, a grining mouth set awry, furnished with long teeth, round the eyes circles of bright brass, surrounded by a larger circle of white paint, from his forehead hung long tresses of buffaloes hair, and from the catch part of his head ropes made of the plated husks of Indian corn." Such a man was most likely a member of the False Faces, an Iroquois healing society, and his performance before (and evidently directed at) Bartram and his companions may have been intended to offset any malevolent effects of their visit. And, apparently, they increasingly elaborated their healing ceremonies, adding the carved masks Bartram described. Archeologists have found no evidence of such masks before the onslaught of epidemics, and such masks as have been found grew successively more grotesque over time. The contorted features of the masks echo the horror of smallpox, a disease in which the entire body is covered with pustules that ooze and stink. But these Indian practices held little interest for colonists, for they saw nothing there relevant to reading God's will.[18]

Everyone could read the message in such unpredictable anomalies as comets and epidemics, for they were always a call to repent and reform, a message applicable to everyone. By contrast, eclipses of the sun and moon, and particularly significant alignments of the planets, could be predicted and interpreted by skillful people, and had more specific messages. An eclipse in 1704, wrote Daniel Leeds, "falling in Gemini . . . denotes Pestilential Diseases, Incursions of Enemies, Death of some Eminent great Man, Tempestuous Weather blowing down Houses, etc . . ." The predictability of these events did not lessen their spiritual significance, for, just like all other natural events, none had any powers that operated independently of God. In the seventeenth century Protestant astrologers explicitly aligned their astrology with providentialism. "God rules this world by Stars," wrote Daniel Leeds in the poem that opened this chapter, and in another almanac he explained, "For though the stars are second Cause of Effects upon Mankind," the first cause was always God.[19]

Congruent with both their sense that their predictions were merely good approximations and an aversion to claiming to know God's will, Daniel Leeds and his son were somewhat diffident about their predictions. They consistently saw it their duty to offer their almanac readers foreknowledge of such celestial events, but made only modest claims for the applicability of their readings. It "is not for Man to be positive," wrote Titan Leeds in 1722. "We may modestly hint at Things, and make our rational Conjectures; for the Stars and the Planets are but Ministers, not Masters; and the Great God Almighty conceals and reserves, some-times, many secret Things from our Fore-sight, which the greatest Art or Experience of Man can never be able to discover, unless by Divine Revelation." In any case, to foresee was not to escape. In 1704 Daniel Leeds offered this comfort:

> If God Protect, no Malice can offend,
> Without his help there's nothing can defend.

When John Clapp failed to predict when eclipses would occur in his 1697 almanac, it was not merely that he demonstrated less skill. If he couldn't figure out when an eclipse would happen, he couldn't help out his fellow New Yorkers by forewarning them about what the eclipses meant.[20]

By contrast, natal or judicial astrology—predicting the personal fortune of a person (or a ship or a city) from the arrangement of the heavens at the hour of birth—is almost invisible in the New York records. This may be in part attributable simply to the paucity of people with the skills to cast natal charts. But more importantly, during the English Civil War in the middle of the seventeenth century, making predictions about the fortunes of particular persons was precisely what brought astrology into disrepute. When astrologers predicted dire effects of celestial events as solely applicable to their political enemies, their credibility was eroded. With the Stuart restoration, such use of astrology was successfully associated with fomenting civil turmoil and was banned. Daniel Leeds was well aware of the hostility towards this aspect of astrology. On rare occasions he or his son did comment that a particular eclipse or a conjunction of planets would affect some people rather than others. But even when a celestial event "causeth the Wrath of Great Men" or would lead to the "Imprisonment of Clergy Men" or affected some similar category of people, Daniel and Titan Leeds hastened to say that they meant no one in particular. Nevertheless, natal astrology was sufficiently familiar as an idea, if not as a practice, for Daniel Leeds to be able to refer to it and assume he was understood. Ever angry with Quakers, he could use Philadelphia's natal chart (metaphorically or not) to declare, "Hence Astrologers judge, That a Town or City founded on such a Configuration will be Prodigious, Commanding and Oppressive."[21]

Everywhere colonists looked, then, God exercised his power in the natural world. Illness came to sinner and devout alike as a reminder of everyone's inherent sinfulness, and whatever efforts one made to preserve or recover health, ultimately healing was in the hands of God. God manifested his displeasure with particular sinners, tested members of his flock, and rewarded his faithful through events that only appeared to have immediate human or natural causes. He sent storms, floods, comets, and epidemics to sweep down on communities at moments of widespread departure from the path of righteousness, warning everyone to repent and reform. Eclipses of the sun and moon, and particular alignments of planets with each other similarly worked his will upon those he wished to chastise or reward. Even if people could and should be active on their own behalf—eating moderately to keep their humors in balance, enforcing quarantines when smallpox threatened, planting seeds when the signs were most favorable—nonetheless, as Stephanus van Cortlandt wrote his daughter, "We are all in the hands of the Lord; His will be done."[22]

God, however, was not the only source of spiritual force active in the world. The Devil, for Christians the incarnation of sin, was also capable of exerting influence, albeit less commonly. By and large, the Devil did not work his will through manifestations in the natural world but rather through affecting human actions. In 1692 jurors in West Chester charged one Gabrill Legett, "having not the face of god before his Eyes, & by the Instegation of ye devill," with stealing a hog. And as Jaspar Dankers traveled to Albany, he encountered a man who admitted having been tempted by the Devil. Consequently, the man had a swearing wife, nasty children, and had experienced one misfortune after another. As the Reverend Megapolensis pointed out when he reported the jailing of the two disruptive Quaker women, "We perceive from this circumstance that the Devil is the same everywhere."[23]

Nevertheless, the Devil was a serious threat to more than those individuals he seduced if his followers became witches, for witches could wreak havoc with other people's bodies, livestock, and crops. Belief in the possibility of witchcraft was widespread among all social groups in early modern Europe, and witchcraft was explicitly criminalized in the Netherlands, England, Germany, and Scotland, the places where most colonists originated. Witches were prosecuted and executed throughout these areas from the middle of the sixteenth century to the last decades of the seventeenth. For Europeans, witchcraft had two aspects: making a covenant with the Devil, in return for which one expected pleasures denied one in a Christian community, and maleficium, the ability to harm others through charms and spells. Fears of people making covenants with the Devil were generally more salient among Calvinist Protestants, for the core of their theology was the covenant with God and so a covenant with the Devil struck at the heart of their piety. Places in which the godly community expended enormous effort to make itself one with the political community by trying to banish the unconverted, like Scotland and New England, were places in which witchcraft prosecutions were notably more prominent. By contrast, places in which Calvinists had less hope of excluding all but covenanted communities, such as England, France, and the Netherlands, were places in which witchcraft accusations were more occasional. Indeed, in the Netherlands in particular in the seventeenth century, skepticism about the reliability of particular claims of witchcraft became more widespread, even as most acknowledged that the Devil was still generally active.[24]

As one might expect, then, such a heterogeneous colony generated only a handful of witchcraft accusations. The one moment when accusations of consorting with the Devil flew heatedly was the period in which Leisler governed. The Leislerians were animated by a passionate rejection of the Catholic-friendly Stuarts, a hatred of popish French Canada, and a hope that they could establish in New York a polity in which Calvinism would be joined with the state. Both they and their opponents claimed to bear the standard of true Protestantism and thus each accused the other of bad faith and leaguing with the Devil. Nevertheless, Leisler and his lieutenant were hung for treason, not witchcraft, for the new government of William and Mary, while committed to anti-Catholicism, did not wish to inflame religious contentions but rather wished to subdue them.

Aside from Leisler and his enemies, accusations in the colony were all for maleficium, not covenant-making, and they were all handled circumspectly. Unsurprisingly, all of the formal accusations arose in towns settled by migrants from New England, as was the case in the story of Mrs. Howells's bewitchment that opened this chapter. During the course of her illness Mrs. Howells cried out that she saw a neighbor, Elizabeth Garlick, in the corner and "a black thing at ye hither corner." Howells was the daughter of Lion Gardiner, the first settler of the town of East-Hampton, and a man who evidently had had at least one conflict with Elizabeth Garlick's husband, Joshua, a man who also had had contentions with others. Mrs. Howells herself had apparently argued with Mrs. Garlick, calling her "a duble tounged woman." In the wake of Howells's death, neighbors testified to Howells's apparent signs of bewitchment, including her claim of seeing Garlick's spectre and her familiar (the "black thing"), and to other troubling incidents which they now recalled as associated with Elizabeth Garlick. A slave child had died "in a strange manner," as had a colonist's child after Garlick had held it, several mothers had had their breast milk dry up after Garlick had asked for some, an ox had broken its leg, and a health sow had abruptly sickened and died. Natural-appearing phenomena, uneasy East-Hampton residents felt, seemed to point to the Devil's handiwork. At the time of these accusations the eastern half of Long Island fell under the jurisdiction of New England and so Garlick was sent to Hartford for trial. She was acquitted and returned to East-Hampton, to live out her life without again appearing in the records. Whatever the suspicions of her neighbors, the evidence against her was evidently insufficient.[25]

Similarly, Mary Wright of Oyster Bay was accused of witchcraft in 1660 and dispatched to Boston to stand trial. Wright was found guilty of Quakerism rather than witchcraft and was banished. In 1665, with all of Long Island now under the jurisdiction of New York, Governor Richard Nicolls convened a grand jury to hear accusations of witchcraft against Ralph and Mary Hall of Seatallcott. A jury of English, Dutch, and Scots men heard testimony that by "Witchcraft and Sorcery" the Halls had caused the death first of one George Wood and then of his infant child. Noting that both Halls had sworn to their guiltlessness and that no one had appeared in person to testify against them (the depositions were read aloud), the jury decided that, while there was reason to suspect Mary (but not Ralph), there was insufficient evidence to convict her. The governor ordered the husband to post bond for the good behavior of the wife and released them. Five years later, his successor, Governor Benjamin Fletcher, heard the case of Katherine Harrison, a widow who had recently moved to Westchester from Connecticut and was rumored to have practiced witchcraft. He ordered her to return to Connecticut in order to quiet the "Jealousyes and feares" of his subjects. She stalled, however, and when nothing untoward occurred, he let her stay.[26]

Officials in the colony, then, did themselves investigate or arrange for an investigation of witchcraft accusations, but with an eye to public peace and holding to reasonably high standards of proof. During the 1692 witchcraft trials in New England, Cotton Mather wrote a number of Dutch and French ministers in New York for their opinions. All readily agreed to the existence of witchcraft, but all dissented from the levels of proof accepted by the New England ministers. In New York, the middling sorts who sat on the jury for the Halls were similarly willing to reserve judgment, even when they found the accusations at least plausible. And at least some spells and charms, things that easily shaded into witchcraft, most probably never came to the attention of the authorities. Poppets, doll-like objects used in spells, have been found hidden in the walls of a house on Long Island built around 1659 whose inhabitants, apparently, never earned the enmity of their neighbors and so never were publicly accused.[27]

When witchcraft accusations did become public, the precipitant was speech as much as deeds, for in the face-to-face world of the early modern Atlantic, speech was experienced as particularly powerful. The salience of witchcraft accusations in New England was at least partially the outcome of the difficulty of sustaining perpetual amity within what were supposed to be communities of saints. But even in New Netherland and New York, where no such universal fellowship was assumed, harsh and angry speech was experienced as dangerous. People who publicly criticized the government, like Henry Smith, who made "idle charges" against local magistrates, or William Taylor, who called Governor Andros "a Rogue and a traitor" while drinking in a tavern, were fined or jailed, for their speech alone was enough to undermine public order. A woman in Elizabeth Town, New Jersey, when that area was under the jurisdiction of New York, was hauled before the court for calling neighbors "witches and devills Whores and Rogues all together." It was as revenge for harsh speech that Mrs. Howells claimed Elizabeth Garlick bewitched her. Howells's mother, in turn, was frightened by her daughter's accusation of Garlick and tried to hush her, saying "[Y]ou are a sleepe or a Dreamed." But Howells's accusation couldn't be unsaid. Speech both caused and reflected the rifts in a community that failed to sufficiently adhere to God, a community in which the possibility of aligning oneself with the Devil seemed always there.[28]

As elsewhere in the early modern Atlantic world, women were regarded as more likely to covenant with the Devil than were men. New Netherland and New York were typical in that four of the five colonists formally investigated for witchcraft were women. Women's intercourse with the Devil was particularly threatening because their compliance with subordination within the household was an important measure of social harmony, and because in their bearing and suckling of infants they produced either an increase in the godly or its opposite. Both of these convictions were particularly strong among the New England migrants, but they were also familiar to other colonists. Daniel Leeds railed against "Children generated of the seed of Deceit, brought forth out of the Womb of Wickedness, and nourished up at the Breasts of Witch-craft, and rock't in the Cradle of Idleness." Leeds was writing metaphorically, but the intense physicality of his images invoked actual women's bodies. The accusations against Elizabeth Garlick included acts like drying up breast milk and killing children, accusations typical of those against suspected witches everywhere in the Atlantic world, acts that were profoundly anti-maternal. Mrs. Howells was herself the mother of a newborn. Her death therefore deprived the infant of suckle and so implicitly threatened its survival. Consorting with the Devil struck at the heart of the ability of colonists to make a social body that was pleasing to God and to sustain that body through their children.[29]

If colonists only saw each other as occasionally tempted by the Devil, they universally ascribed Devil worship to local Indians. "They worship and present offerings to the Devil," declared Johannes Megapolensis of the Mohawks in the 1630s, repeating an assumption widespread in early modern European accounts of New World natives. "[T]heir worship . . . is diabolical," concurred Daniel Denton of the Algonquians of Long Island. Harmen van den Bogaert described how a group of Oneidas had invited him to watch them "drive out the devil" from an ailing man. William Hyde, posted to the Albany and Schenectady forts in 1696, described watching an Iroquois ceremony in which a "Conjuror" led a group in "such a frightfull noise among them Like the yell of Wolves, Till the Devil appears to them in the shape of a Beare, Woolfe, or fox, Catt or some other Wild Creature, which as soone as vanishes." Then, Hyde recounted, "The Old Wizard Rises and tells them what success they are to have."[30]

And in the eyes of the colonists, the relations of the Algonquians and the Iroquois with the Devil allowed them to engage in magical practices beyond healing. Van den Bogaert described being shown objects made from animals and other materials that the Oneidas kept for "idols or telling fortunes." And Wentworth Greenlaugh, traveling in Iroquoia in the 1670s during a period of warfare between the Iroquois and the French, recalled being awakened during the night by "a great noyse." It was, he reported, the Cayugas driving away the ghosts of the captives they had just ceremonially killed.[31]

Colonists were also sometimes aware that the Africans or African-descended people they held in slavery practiced magic. In the 1712 rebellion, the conspirators pledged themselves to secrecy "by Sucking ye blood of each Others hands." And when the whites counterattacked, the slaves at first believed themselves invulnerable because "a free negroe who pretends to Sorcery gave them a powder to rub on their Cloths." As the minister John Sharpe pointed out, the leaders were from the Akan-speaking Gold Coast, an area rich in magical practices and complex spirituality. Many of the slaves who labored in New York had grown up free in Africa, and consequently could bring with them the spiritual and ritual practices of their homelands.[32]

While white colonists occasionally reported the magical practices of Indians or Africans, they expressed little concern about them. Sometimes they simply dismissed the magic of Indians and Africans as ineffectual. The conjuror's powder did not save the rebelling slaves, and occasionally a colonist mocked the Indians' practices as merely delusional. But by and large, they recounted these stories as if the Devil was indeed a participant. Nevertheless, any consorting with the Devil that Indians or Africans might do seemed to have no potential for threatening colonists, unlike when one of their own became a witch. Colonists in the seventeenth century never asserted that Indians or Africans were unalterably different from themselves. Indeed, the reluctance to allow African conversions stemmed from Africans' potential to become genuinely Christian. Nonetheless, despite their contentions with one another, colonists believed that they formed a single social body, one to which Indians and Africans were outsiders. Devil-allied colonists could wreak havoc with other colonists' children and livestock, but any magic Indians and Africans wrought affected only themselves.[33]

*I*f natural disasters and witchcraft spoke unambiguously of God's judgment and the Devil's wiles, there were nonetheless other anomalies that spoke less dreadfully but also less clearly. These more minor unusual occurrences were sometimes called "prodigies," sometimes "curiosities," but more generally "wonders." Wonders were of interest throughout the early modern European world, in New Netherland and New York no less than elsewhere. Wonders titillated and entertained, and they offered opportunities for sociability. While they also simultaneously affirmed beliefs about the nature of the natural world, they also revealed that, while colonists had a general consensus on the significance of spiritual forces in the material world, this general consensus did not mean uniformity.[34]

The most common wonders were exaggerations of the everyday: animals from exotic places, misshapen animals, human dwarves and giants, conjoined twins. When William Bradford began to publish a weekly newspaper in New York City in 1724, he reported such wonders from elsewhere. Settlers could read of a "Sea-Lyon" killed in Boston and an eight-legged kitten born in Philadelphia. Sometimes New Yorkers could see such wonders for themselves, as they could in 1728 when a lion was exhibited at a fair on Long Island. The lion evidently was toured around, for a farmer in Connecticut recorded viewing it and noted in his diary that it had already been shown in New Jersey, New York City, and Albany, as well as on Long Island.[35]

By and large, these reports were brief and the significance of these oddities was unstated. Upon occasion, however, an event was deemed worthy of more extensive analysis. In 1732 a woman gardening in New York City pulled up a radish and "found that the Stem of the Radish grew out of the Appearance of a Child's Hand and Fingers." This being "surprisingly strange," it was taken to a magistrate who ordered it preserved in alcohol. The radish was put on display and, the *New-York Gazette* reported, "continues in the Shape and Colour of a humane Hand and Five Fingers with Sinews and Joynts which open and shut." An "Abundance of People resort daily to see it," and they speculated that an infant had been buried in the garden. According to the newspaper, New Yorkers guessed that the radish seed had taken root in the wrist of the child and "the Vegetative Quality of the Radish . . . preserv'd the Flesh from putrifying." Here plant and human interpenetrated, the permeable borders of the two sorts of material substances being no barrier to their interaction. And here too the evidence of the material world was of the human world gone morally awry, for surely the child must have been misbegotten, for why else would someone have buried it in unhallowed ground?[36]

The *Gazette* story is the single surviving account of the anomalous radish and may well have reported more of a consensus than in fact existed. Sharing a common Protestantism, there seems to have been no significant differences among colonists in New Netherland and New York in their willingness to recognize the hand of Providence in natural events. The propensity to see the effect of witchcraft in anomalies seems to have especially characterized New England migrants on Long Island and near Connecticut, even if everyone assented to the existence of the Devil. Wonders less rooted in the Bible, however, were more open to interpretation. An anomaly with several surviving accounts shows that people in different social circumstances tended to interpret these sorts of oddities differently. As described in the introduction, a giant tooth was unearthed in Claverack, a town on the Hudson south of Albany, followed by more teeth and bones. The Dutch colonist who brought the first tooth to New York City and showed it around thought of it as a simple amusement, "worth nothing." Governor Cornbury, who sent it to the Royal Society in London, thought of it as something worthy of investigation by those interested in natural history. Local River Indians understood the bones and teeth as part of their own history, the remains of a giant person about whom their fathers had told them. And when Governor Joseph Dudley of Massachusetts saw one of the other teeth, he found confirmation of the biblical Flood. Here the variety of interpretations of the significance of these remains followed the configurations of social differences. Governor Cornbury, embedded in empire-wide networks of patronage, saw them as something that would interest his peers in London, men engaged in collecting cabinets of curiosities and in learned discussions of them. Governor Dudley assimilated them to Christian sacred history and used them as an occasion to contemplate biblical revelation. River Indians asserted their own ownership of the meaning of the remains, while more modest New Yorkers seemingly gave the remains little significance but were happy to give or trade them away. While some natural phenomena, then—notably those with clear precedents in the Bible—were understood similarly by most colonists, others allowed for variations in meaning that stemmed from different social experiences. But note also that no one seemed to particularly have exerted themselves to convince the others that their interpretation was true. Wonders allowed people to make of them what they might.[37]

$\mathcal{A}$s we have seen in part 1, then, understandings of the natural world common among colonists in the seventeenth and early eighteenth centuries were drawn from those common in northwestern Europe. They could do so because their familiar agricultural practices, their crops, and their farm animals transplanted relatively easily to Long Island and the Hudson Valley. Their common explanations for the everyday—the circulation and corruption of humors in human and animal alike, the passage of moisture, temperature, and toxins through the skin, the relative openness and sweetness of soil, the shifts in the terrestrial world that followed shifts in the celestial one—assumed a natural world that was everywhere the same. Honed through interactions with the material world, their knowledge was consistently confirmed by bodily experience, what each person saw, felt, and did for himself or herself. Anomalous events like comets and epidemics were as easily explained here as they were in Europe, for God was an active presence, shaping their lives through his Providence and chastising them for their sinfulness. And, as in Europe, the Devil was dangerous, tempting an occasional colonist into wayward behavior or even into witchcraft. Insulated by their technology, by their conviction that they understood the natural world, and by their spiritual beliefs, colonists in New Netherland and New York had to master relatively little knowledge of what was peculiarly local and had to adapt their practices only modestly.

What adaptations they made that accommodated to the natural world that they found in the region were modest extensions of what they already knew and did. As was true everywhere in the early modern Atlantic world, labor that depended upon local soils and waters, plants and animals, and weather and seasons determined peoples' fates. Their work in the natural world was intended both to sustain themselves and to enhance their fortunes. Whether the objective was beaver, wheat, whale oil, or salt meat, knowledge and skills concerning the natural world were directed toward the extraction of commodities and the expansion of trade. The technology they brought with them—the plows and cattle, printed advice and written contracts, ships and mills—enabled them in large degree to ignore the subtle idiosyncrasies of this particular natural world. Even if their livestock foraged freely, they followed Indian trails, and they encountered Akan conjurors, colonists in New Netherland and New York primarily attended to what was familiar and expected.

The economy of knowledge followed the social composition of the colony. Absent universities, cities, and courts, with few of either the erudite or those with wealth enough to patronize them, the beliefs and practices of the middling sorts dominate the records. With the caveat that we simply know little about what those at the bottom—slaves, sailors, servants—knew and did regarding the natural world, the people who published almanacs, served on coroner's juries, read *The Husband-man's Guide*, delivered babies, surveyed land, and wrote spouses, parents, and children about the daily life of health and agriculture lived in a common natural world. This was a place in which almost everyone labored with crops and livestock, with ailing bodies, with winds and tides, and saw God's hand everywhere. Consistently confirmed by the evidence of experience and of their own senses, their beliefs and practices allowed for variation, for everyone's experiences were different, expectations were modest, and approximation would do. Less elaborate and intellectually consistent, perhaps, than systematic efforts to apprehend the natural world among the learned in Europe, in New Netherland and New York the beliefs and practices of the middling sort were the bedrock of everyday life.

PART II

New York as a
Province of Britain,
1720–1775

CHAPTER 4

Improvement

The Newspaper

'Tis truth (with deference to the college)
News-papers are the spring of knowledge,
The general source throughout the nation,
Of every modern conversation. . . .
For those who are but puny made
Are arts and sciences and trade . . .
If you want health, consult our pages,
You shall be well, and live for ages . . .
Our services you can't express,
The good we do you hardly guess;
There's not a want of humankind,
But we a remedy can find.
New-York Weekly Post-Boy, *16 April 1770*

On a June day in 1714 a crowd gathered along the shores of Manhattan to watch a novel sight. On this particular day, the throng included local notables, like members of the General Assembly and the governor, Robert Hunter. The object of their attention was one Joseph Morgan, who was maneuvering a rowboat just offshore. But this was no ordinary rowboat. Morgan had fitted up the boat with wheels, springs, cranks, and weights, a system all of his own invention. Ordinary rowboats, although fairly simple to build and easy to learn to use, were subject to the power of tides, winds, and waves, and to the limits of human muscle. Morgan hoped his rowboat would overcome those impediments and thus be an "improvement."[1]

In the eighteenth century, time's arrow changed. Before roughly 1700, ordinary New Yorkers, like most people in the early modern world, believed themselves to be living in a time of the decline of knowledge. At the dawn of time people had had perfect knowledge, and even after some lapse of time, the ancients still retained far more of that perfect and complete knowledge than people now did. Time's arrow pointed backwards, toward Eden. But in the decades after 1700, time's arrow wheeled around, pointing toward the future. New Yorkers began to believe that they were living in a time of the progressive unfolding of new understandings, an era of discoveries. In 1736 one New Yorker wrote that it was in fact the past that had been "a Green headed Time," one in which "every useful Improvement was hid." In 1765 Roger More printed a special chronology in his almanac. The list began in 1208 and, running up to the present, included such notable improvements as the invention of spectacles and the introduction of turnips. In this account, improvement was cumulative and led into an ever-rosier future.[2]

Improvement was, of course, the primary reason for colonization, any colonization. European polities from Portugal to the Netherlands sought colonies because the exploitation of distant places mobilized commodities that improved their wealth. Kingdoms like England and Spain rewarded men with land grants in distant places because successfully holding territory improved the political power of the realm against their rivals. And, on a personal level, individuals moved out to colonies to seek their own improvement: their material betterment, their social elevation, or simply their escape from the law or unhappy circumstance. New York was no different. Both the formal institutions of empire, like the Dutch West India Company and the British Crown, and the immigrants from Europe sought the improvement of their fortunes through the colonization of the region.

Nevertheless, in the eighteenth century improvement developed a prominence it had not had in previous centuries. Like people elsewhere, New Yorkers came to have material experiences of specific innovations that softened, somewhat, their sense of vulnerability to the natural world and made them feel more in charge of their destinies. The natural world was slowly seen to be more reliable, and people became somewhat more confident that their efforts in it would bear fruit. As settlements in New York grew more contiguous, filling in the local landscape, the vastness of the wilderness began to recede to the horizon. Through an ever-increasing volume of print, they were increasingly aware of a variety of inventions in the wider world. In the 1720s New Yorkers got their first weekly newspaper, got a second in the 1730s, and from the 1740s on always had at least three papers, all bearing news of inventions and innovations. Almanacs proliferated, and they expanded the sorts of advice Daniel Leeds had given his readers. By the 1750s booksellers offered a host of the kinds of self-help books that allowed people to improve themselves. Over the eighteenth century, New Yorkers grew a wider variety of crops and engaged in a wider variety of crafts. Improvements to navigation, along with a rising population in the colony, allowed for networks of trade to become stronger, more extensive, and somewhat more reliably lucrative. In short, through experience, improvement became associated less with outwitting the vagaries of fortune and more with commanding success.[3]

And yet, all places have their idiosyncrasies and New Yorkers adopted and adapted improvements that circulated in the wider Atlantic world in particularly local ways. Here as elsewhere, the path to improvement was not smooth. Instead, improvement as both a program and a practice was bound up with social relationships—relationships between the prosperous and the struggling, between the colony and Britain, and among neighbors, friends, kin, and acquaintances that ran the gamut from trust to hostility. Consequently, there was no simple relationship of cause and effect between the introduction of an innovation and its local adoption; people rarely, if ever, adopted something simply because it worked. The willingness to adopt new things or methods was as much a function of social relations, of who advocated something, as it was of its material nature, what it was.

In New York, some areas of improvement, like inventions, were less marked than elsewhere, some, like smallpox inoculation, were more, and others, like improvements to agriculture, were very similar to the pattern found elsewhere in the Atlantic world. Actual inventiveness was episodic and never caught on as a notably admired or emulated practice. Inventions developed elsewhere were adopted readily enough when they seemed commercially viable, but for reasons that are not altogether clear, few New Yorkers joined Joseph Morgan in seeking mechanical improvements.

Inoculation for smallpox, by contrast, became remarkably well established in New York, as much, if not more, than anywhere else in the Atlantic world. Inoculation was indeed a genuinely useful response to smallpox. However, as we shall see, its adoption in New York did not follow straightforwardly upon the recognition of its efficacy, but took a more circuitous route. Crucially, its eventual acceptance depended upon social relations in the colony: who adopted it and how it was promoted.

Agricultural innovations were familiar to New Yorkers, and were indeed adopted steadily and widely. But this was also an area that was bound up with social status, with what marked a "gentleman" from the common sort. In the 1760s a group of such self-declared gentlemen actively tried to encourage agricultural improvement in the colony, an effort by the local elite to speed the shift toward the embrace of improvement. Their improvement society was a spectacular failure, not because people of lesser status particularly resisted such tutoring but because this attempt at an alliance directed towards the public good crashed on the rocks of private rivalry. The failure of these men to demonstrate their ability to foster the public good through such a gentlemanly realm as an improvement society reinforced more ordinary people's distrust of their claims to political leadership. While the embrace of improvement by the 1770s came to characterize New Yorkers, the particular configuration of the embrace depended as much upon people's relationships with each other as the nature of the improvements themselves.

*T*he story of Joseph Morgan demonstrates both the appeal of inventing and the weakness of the embrace of that particular form of improvement in colonial New York. Born in 1671 in rural Connecticut, Morgan was ordained a minister in 1699 and then earned his BA from Yale in 1702. Subsequently he served in churches in Greenwich, Connecticut, and in Westchester County, New York, before settling permanently in Freehold, New Jersey, in 1710. He lived his adult life, then, in a geographic arc about a day's ride or so from New York City but nonetheless largely among modest farm people.[4]

Like most ministers, he had to supplement his salary with farm work, but he also had experience as a miller, work that apparently nurtured his mechanical interests. The problem of inland navigation was a particular concern of his. Besides his mechanical rowboat, he drew up proposals for other navigational aids: a method for dredging deeper channels in shallow rivers, a system of sluices and locks for circumventing falls, a winch for hoisting boats over portages, another boat designed for going up swift streams, and yet a third boat, a sort of combination raft and canoe, intended to transport goods on lakes.[5]

Morgan did not merely tinker. In 1715 William Bradford published a novel Morgan had written, *The History of the Kingdom of Basaruah*, an allegory of a man's fall and Christian redemption. Morgan also published a number of his sermons and other short theological pieces. While these productions might seem to us to be of a different ilk than diagrams for locks and boats, they were not so disparate to him. Morgan did not take credit for his inventions but rather described himself as "him that Almighty God has made the Inventor." His inventions, then, were like Dr. Nicoll's medical advice in chapter 3: successful only with God's aid and blessing. And he did not imagine his inventions as merely useful to his own welfare. Like the concerns of his novel and his theological work, in which he laid out programs by which people could bring about, as the historian Ruth Bloch notes, "Christ's Kingdom in the New World," his inventions were intended to help ordinary colonists thrive in peace and prosperity. Such inventions might change the material world for the better, but Morgan did not imagine a rupture with a world in which Providence reigned. If the vigor with which he pursued improvements was novel, the novelty lay more in the hopefulness he expressed, compared with the submission to adversity expressed by someone like Maria van Rensselaer, than in objectives, for both wished to live in accord with God's will.[6]

Morgan did something else unusual with his productions: he sent them to the Royal Society of London. After his rowboat demonstration, Morgan sent a diagram of the boat to them with a letter of explanation. The Royal Society, he hoped, would find this improvement on an ordinary longboat suitable for the Royal Navy and would show their appreciation for his invention by giving him a suitable reward. He also sent them a proposal for "A Method to Oblige the Inhabitants of North America to forever [illegible] Loyal to the Crown of Great Britain." The gist of this method was "Let us have no great Men." Rather, estates should be divided equally among sons, and new settlements established based on covenants like those in Rhode Island and Connecticut. The middling sort, owning their own property, governing themselves, and pursuing their own prosperity, with no great landowners, merchants, or governors, would stoutly support both the king in Britain and God in heaven, thought Morgan. Rowboats and canal locks were like property division and covenants: instruments for the social and spiritual improvement of mankind.[7]

That Morgan should send all his ideas for improvements to the Royal Society shows how this institution had become widely recognized as the center of innovative knowledge about the natural world by the beginning of the eighteenth century. Established in 1662, just before England took New Netherland, the Royal Society had a brilliant coterie of members in the last third of the seventeenth century, men like Robert Boyle, Robert Hooke, and Isaac Newton. By the time Morgan was a student at Yale, the Royal Society was the hub of a network of people interested in the natural world, centered in London but spread throughout Britain and Britain's colonial possessions, and with connections to the like-minded throughout Europe. At its meetings members read aloud letters and reports on curious phenomena from all over the globe, examined specimens presented by members and correspondents, watched experiments, and occasionally financed expeditions. They published what they judged to be the most valuable of these in their annual *Philosophical Transactions*. It was to the Royal Society that Governor Lord Cornbury dispatched the giant tooth from Claverack in 1705. In 1714, when

Morgan plied his mechanical boat off the shores of New York, Governor Hunter was himself a Fellow of the Royal Society, and indeed was the route by which Morgan sent in his rowboat proposal. Long after Hunter left his post, Morgan would continue to send his ideas to the Royal Society, writing letters right through 1739.[8]

No one ever answered. Of all his ideas, Morgan only made a prototype of the boat he rowed off New York because, he explained, his income mostly went "to maintain a great Family of Small children." Neither money nor encouragement was forthcoming from London. At one point Morgan wrote that he assumed his letters had gone astray, but, he added, it would be kind if someone there could let him know. But London was too far and Morgan too obscure. With the best of wills he could not create a network with people so distant and so unknown.[9]

And few people around him seemed much interested either. At one point Morgan conveyed his proposal to make shallow rivers navigable to Benjamin Franklin, who published it in his *Pennsylvania Gazette* in 1732. As we all know, Franklin was luckier, for he had a group of friends who became his fellow tinkerers and speculators. While Franklin became the most renowned, without the conversation and shared activities of men like Ebenezer Kinnersley and Thomas Godfrey, he would likely have been just a printer. But Morgan never found such fellowship in rural New Jersey, nor were there many men in nearby New York who ever developed such interests. An occasional New Yorker made a stab at inventing some improvement. In the 1730s Lewis Morris evidently used a plow of his own invention. Cadwallader Colden, a Scot who moved to New York in 1719 and whom we shall encounter again, sent proposals for an improved quadrant and a new method of printing to London in the 1740s, neither of which, apparently, was workable. One S. Deane advertised in 1766 that he had discovered a method for making tidal mills work throughout the tide's cycle, which he would demonstrate for a fee, but there is no record of any outcome. But these were isolated endeavors.[10]

For historians, it is hard to analyze absence, to figure out why no culture of invention developed in colonial New York. Why didn't others gather around Morgan or Morris, or try to better Colden's inventions or Deane's? Certainly New Yorkers could read about the increasing pace of invention elsewhere. Occasionally up to the 1760s, and thereafter with increasing frequency, New York newspapers reported inventions like a fire grate that decreased the smoke in rooms, a method of desalinizing sea water, and a canvas tunic with corks for sailors swept overboard in storms. Other articles

reported on new canal building and other improvements to inland navigation. By the middle decades of the century, some of these improvements began to appear in New York itself. Paper mills, for example, joined the gristmills and sawmills long established in the colony. In 1745 London-trained
Anthony Lamb opened an instrument shop that offered improved sextants,
compasses, and other instruments for surveyors and sea captains. Peter Hansenclever, a German who arrived in 1764 with money to invest on behalf of
wealthy Britons, bought a struggling old iron works in northern New Jersey
and revived it with new techniques. But while New Yorkers increasingly
benefited from technological innovations developed elsewhere, few of them
were ever drawn to innovate themselves. However much improvement got
good press, in the realm of inventions New Yorkers seemed largely content
to adopt what others did.[11]

*W*hile New Yorkers were apparently generally indifferent to the
delights of inventing, they adopted inoculation for smallpox as rapidly as
did any group anywhere in the Atlantic world. Unlike mechanical inventions, however, inoculation was controversial. Indeed, in some places inoculation was resisted into the nineteenth century. By 1700 smallpox had
emerged as the most deadly of the epidemics that circulated in the Atlantic
world. And over the course of the eighteenth century, as the population
density of the Atlantic rim rose, smallpox epidemics became more and
more frequent.

Smallpox was a particularly dreaded experience. As people throughout
the Atlantic world well knew, smallpox primarily spread through person-to-
person contact, although it could also spread via dried exudates on bedding
and clothing. After a symptomless period of ten to fourteen days after contact with an infected person (a characteristic that allowed people to spread
it before they were aware they were ill), the first sign of smallpox was fever
and pain. Four days later fluid-filled pox erupted through the skin and in
the mouth and throat. The pox made it excruciatingly painful to swallow
and the pox-covered skin stank like rotting flesh. In the worst cases, people
died before the pox even erupted, their bodies turning purple from blood
vessels rupturing beneath the skin, or they suffered for weeks before dying,
the eruptions of pox so massive that they all ran together, the skin peeling
from the body. The mortality of these epidemics varied from episode to
episode, sometimes reaching 30 percent or more, a fearsome toll. If the
person survived, the eruption of pox would taper off, with scabs forming
after fifteen to twenty days. The survivors were sometimes left blind by

pox-scarred corneas, pregnant women often miscarried, men sometimes found themselves sterile, and all of them were disfigured.[12]

If smallpox sufferers lived, however, they were immune to the disease for the rest of their lives. Inoculation was a procedure that was intended to give people cases of smallpox that were mild enough so they would likely live through it but that would render them immune. The procedure entailed putting exudate from an active pox into an incision in the arm or leg. It caused a genuine case of smallpox, but for reasons that are still not understood, almost always resulted in a mild case. The mortality was indeed low, perhaps only 1 to 3 percent.[13]

Information about inoculation entered the Atlantic world initially from the Ottoman Empire and West Africa. Around 1700, members of the Royal Society in London heard about the practice as it was done in Constantinople. In 1714 the *Philosophical Transactions* published a description of it written by Emanuel Timonius, a physician trained in Padua and Oxford who was working in the Ottoman capital. In 1721 Lady Mary Wortley Montagu, the wife of the British ambassador to the Ottoman Empire, returned to London and had her child successfully inoculated. Subsequent well-publicized trials in London had similarly happy outcomes. Meanwhile, in Massachusetts the minister Cotton Mather, who had wide-ranging interests in the natural world, first heard of inoculation in 1706 from a slave from West Africa and then read Timonius's 1714 account. When smallpox reappeared in Boston in 1721, Mather and a group of doctors promoted inoculation. Other doctors and a group of lay allies, however, opposed this novel procedure and there ensued a furious war carried on in pamphlets and newspapers.[14]

It is hard to know, however, how many New Yorkers were aware of these events. During the Boston controversy, Cadwallader Colden commented on it in a letter to a friend there, so apparently news of it had reached New York somehow, whether from travelers, by private letter, or through the sale of the Boston inoculation pamphlets. In 1728 Bradford's *New-York Gazette* printed a short item reporting that "Inoculation has been perform'd [in England] with universal success by Dr. Elderton of Salisbury"; since the item offered no explanation of what inoculation was, the editor must have assumed that his readers had heard of it. But in the 1720s no smallpox appeared in New York, and so the absence of conversation about it is unsurprising.[15]

Then in the summer of 1730 the *Gazette* reported the reappearance of the disease in Boston. The next spring it was in the Mohawk Valley, where the paper reported it was "very Mortal," and by the summer of 1731 smallpox had spread throughout the region. In August the *Gazette* began to report the weekly tallies of the dead in New York City. Few homes were free of it, and one-third of those who got it across the river on Long Island reportedly died. In September, as the death toll rose, the *Gazette* reprinted the description of inoculation that had been printed in the *Philosophical Transactions.*[16]

"Inoculation takes mightily on Long Island," James Alexander, the man who had tried to establish the New York–New Jersey boundary line in 1715, wrote Cadwallader Colden that December. Alexander reported having been told that at least seventy people there had had themselves inoculated in the past two weeks, and fifty more intended to be inoculated the next week. He had heard that others had been inoculated in New York City and in Morrisania, north of the city. In January the *Gazette* reported that inoculations were being done in New Jersey. In March 1732 Alexander wrote Colden that people continued inoculating themselves on Long Island and that it had had "Success beyond Expectation."[17]

Here, then, appears to be the perfect circumstances encouraging the embrace of an improvement: an urgent threat, a set of clear directions, and the concrete experience of success. But while in retrospect we know that inoculation worked, for people in the eighteenth-century Atlantic world, who faced repeated threats from horrific epidemics and an unfamiliar procedure from foreign places, it was initially not so clear. These people had through hard experience learned to respond to epidemic disease with quarantine and isolation, and inviting smallpox to come among them via inoculation must have seemed bizarre. And indeed, smallpox returned to New York in 1738 and again in 1746, and both times New Yorkers primarily (and not unreasonably) tried to curtail its effects by dramatically cutting back on the circulation of people and goods. In 1739, for example, the colony's General Assembly held its session outside New York City, as a third of the members reportedly had not had smallpox, and in 1746 trade in the city virtually ceased, for "the Small pox keeps the Country people & ye Indians from Town." The alternative of inoculation, by contrast, generated "warm and unchristian Debate," despite some people's experiences of it. However much James Alexander could claim in 1731 that inoculation had been hailed, some New Yorkers were reluctant to accept it, for inoculation was profoundly counterintuitive. Why should the disease when contracted deliberately be milder than when contracted at random? Moreover, New Yorkers continued to believe that each body was idiosyncratic, with its own humoral constitution. In all illnesses, what worked with one person did not necessarily help another. Inoculation as a singular technique, done the same way to one and all, was profoundly at odds with common understandings. Nevertheless, by the time smallpox recurred yet again in 1752, the debate had largely died away and the controversy resolved in favor of the widespread acceptance of inoculation. How did this happen?[18]

Critical to this process of making inoculation acceptable was integrating it into familiar ways of responding to smallpox. People had long recognized that surviving smallpox was related to the way the pox erupted through the skin. One local minister wrote his son that his smallpox-afflicted wife blessedly had little pain. "But alas!" he worried, "the pock does not fill well!" and to his great sorrow, his wife indeed died. In mortal cases the pox either never emerged at all or erupted so violently that they ran all together. Unlike the resolution of other epidemic illnesses, "the dissease cannot be carried off by any other evacuation," commented Colden, "but by eruption of Pustles on the Skin." Caring for the afflicted, then, was directed toward trying to encourage the emergence of the pox to the proper degree and at the proper pace. In the sixteenth and seventeenth centuries, as smallpox loomed ever larger on the disease landscape, Europeans largely followed the advice of the Persian physician Rhazes: they kept smallpox sufferers warm with fires and blankets in closely shut-up rooms, and gave them alcoholic drinks ("cordials," thought to have warming properties) to keep the pores of the body open. In the seventeenth century, the English physician Thomas Sydenham argued for a cooling regimen instead: open windows, fresh air, a light diet, nonalcoholic drinks. In New York, Colden embraced Sydenham's regimen, while farmers near him in the Hudson Valley continued to rely on rum, but all anxiously watched the sequence of the eruption of the pox.[19]

Even if the smallpox characteristically vented itself through the skin, it was nonetheless an ailment of the whole body. Fever, delirium, and the consequences of a general corruption of the humors had to be managed too, and it was here that inoculation was reimagined so that it made sense. Inoculation allowed people to time the onset of smallpox, allowing them first to bring their bodies into an optimal state to receive the infection. While the inoculation directions published for New Yorkers in 1731 merely described the inoculation procedure itself, by the 1750s the focus was as much on the preparation of the body to receive the pox exudate as it was on inserting the exudate itself. Now people being inoculated underwent several weeks of purging, bleeding, and a limited diet, all intended to rid the body of excessive or corrupted humors before smallpox was introduced. And if the insertion of the exudate was the same for everyone, the preparation of the body was tailored to each person's constitution. When Colden, for example, oversaw the inoculation of his son David, a young man in chronically frail health, he commented that David's preparation was necessarily different from that practiced "among the Country People, who are generally of Strong robust constitution." By the 1750s the usefulness of preparing the body to withstand the rigors of smallpox, bringing each person's body into optimal balance, was so well accepted that people began to prepare themselves at times when they recognized that they were at risk of getting smallpox the "natural" way, even when they did not intend to be inoculated.[20]

And controlling the timing also allowed people, to some degree, to undergo smallpox when the surrounding air was least threatening. Spring or fall, when the temperature was moderate, were the best seasons to inoculate, in distinction to winter when cold air threatened to close up the pores or summer's heat that encouraged too hasty and too violent an emergence of the pox. James Alexander attributed the death of a few of the early cases of inoculation to going outside in winter weather, so that the people "by carelessness catcht Cold, which Struck them [the pox] in again." One July another New Yorker marveled that people "do inoculate some even in this hot weather." When inoculation was understood in this light, as a technique that allowed people to bring their bodies into the optimal condition with which to withstand the disease rather than a condition that in itself somehow set the disease off on a distinctively milder course, and to choose the optimal time to undergo the disease, it made sense.[21]

These adaptations, however, were not enough to settle the controversy over inoculation's adoption in New York. In the eighteenth century no less than in the seventeenth, epidemics were interpreted as acts of Providence, instances of God's messages to repent and reform. In ensuring a safer passage through the disease, inoculation seemed "like taking the Prerogative out of the Hand of God—distrusting his Providence," as one person wrote in a newspaper. In the eighteenth century as in the seventeenth, all Protestant denominations in New York continued to affirm God's continued superintendence of the here and now. Smallpox had long been an example of his chastisement that called for repentance and reform, and survival had long been attributed to his blessing. Abruptly, inoculation threatened to drain smallpox of its theological significance. As such, opponents claimed that inoculation was impious.[22]

Emerging in the 1730s and dominant by the 1750s, a different explanation for the root cause of smallpox countered this charge of impiety. In an essay published in the *New-York Gazette* in 1738, an anonymous author asserted that smallpox was "no more than a natural Fermentation of the whole Mass of Blood of an human Body, occasioned by those Foeces in it, which it Originally contracted in the Womb, and which, at one Time or another, most human Bodies throw off, by this means." A supporter of inoculation declared in the *New-York Mercury* in 1760, "I already have the Seeds of the Distemper in my Blood, which wants nothing but the Affluvia to set these at work. Inoculation is only setting those in Motion, in order to a more easy Discharge." Imagining smallpox as innate, rather than a disease that arrived from outside, removed it from the general run of epidemic disease. Smallpox mimicked contagious diseases only because some sort of "Affluvia"—noxious air or matter impregnated with something thrown off by smallpox sufferers—made the disease manifest itself in people nearby. And if it was innate, it was also universal, a disease that could erupt naturally, not as a manifestation of God's particular displeasure. One of several explanations for smallpox that circulated in the eighteenth-century British Atlantic, the innate seed theory appealed to inoculation supporters because it so successfully countered the charge of irreligion. Indeed, by the 1760s proponents were asserting that it was inoculation itself that was providential, that it was "a discovery GOD in his mercy has been pleased to bless mankind with."[23]

Support for inoculation also came from a different direction, and one that was relatively novel: statistics. The useful mathematics—the arithmetic used by merchants and shopkeepers, the geometry and trigonometry used by surveyors and navigators—generally gained ground steadily in New York over the course of the eighteenth century. In 1730 the first mathematics teacher advertised his evening school in New York City, and by the 1760s a young man could study mathematics in at least four such schools there as well as ones in Hackensack, New Jersey, and near Newburgh, up the Hudson. Among the middling sort in New York, numeracy was modest (almanacs, for example, often printed tables of various sorts for those whose arithmetic was weak) and precision still not particularly valued. Inoculation, nonetheless, was an instance where numbers were offered as incontrovertible facts in order to persuade people of its efficacy. In 1738 the *New-York Gazette* reported on inoculation's success in the Carolinas. Not one of more than fifty inoculated whites died, the newspaper reported, and only two or three of the still larger number of inoculated slaves. By contrast, "at least" one in five died among those who had caught smallpox naturally. New York newspapers then printed a spate of such statistical accounts in 1752 and henceforth occasionally printed accounts from other places. The use of numbers in affirming the benefits of inoculation reflected the slow rise of the prestige of calculation, a process in which numbers came to represent "pure fact," a datum of

reality that stood apart from opinion or point of view. Such statistical argu-
mentation is familiar to us, but was just becoming visible in the eighteenth
century Atlantic world. Note that in New York, this marshalling of statistical
evidence became most prominent after the inoculation controversy had died
away. That is, statistics affirmed what people had come to believe on other
grounds, and so may have been an instance in which experience enhanced
the prestige of statistics and vouched for their reliability, rather than the other way round.[24]

Inoculation was incorporated into New Yorkers' repertoire of healing
in the twenty years after its introduction, then, because they could reimagine
it. Because they could focus on the preparation of the body, they successfully
integrated it into familiar ways of treating smallpox. And reconceptualizing
smallpox as innate altered its providential meaning, translocating that meaning from the disease onto inoculation itself. But another important reason
why inoculation was accepted relatively quickly in New York was that it was
practiced by ordinary people.

While conceptually inoculation had to be reimagined in order to make
it congruent with New Yorkers' understanding of the body and illness, it
disturbed ordinary social relations of healing not at all. In the eighteenth
century, healing continued to be part of everyone's repertoire of skills and
care of the sick continued to be centered in the household. In the accounts
of the first inoculations in New York, most took place in towns like Jamaica and Morrisania, towns populated largely by struggling and comfortable
farm families. James Alexander mentioned a man named Van Brunt on Long
Island as having inoculated his family of seventeen. Alexander's wording
indicates that Van Brunt himself did the procedure, and with a "family" of
seventeen, it is likely that Van Brunt was one of the many prosperous Dutch
farmers of King's County who owned slaves. Nonetheless, there is no record
of Van Brunt holding political office or otherwise achieving any prominence.
Since newspapers printed letters and essays generally without attribution, it
is difficult to know the social configuration of those who supported and who
opposed inoculation during the 1738 and 1746 outbreaks, or if there was any
social difference between them. Certainly none of the wealthy men of the
colony stepped forward as a public advocate of either position. What scanty
evidence there is indicates that well-to-do families made private choices
about inoculation just as more modest New Yorkers did. Peter Livingston,
for example, a grandson of Alida and Robert, inoculated his two children,
but the evidence for this comes from a private letter. Wealthy families, as the
Livingstons now were, never publicly pronounced on inoculation one way

or another. Nor did any of the few men in New York with formal medical degrees make public pronouncements.[25]

The one public statement on inoculation by a colonial official was Governor George Clinton's ban on the practice in New York City. But he did this as the 1746 outbreak began to die away in 1747, and not in opposition to inoculation per se, merely its reintroduction into New York City. Inoculated patients were as contagious as were smallpox sufferers who had caught it naturally, but because the patients sometimes were barely ill at all, they felt free to go about their normal business and could thus spread the disease. Clinton said nothing about inoculation elsewhere in the colony, and neither advocated it nor disparaged it in his ban. Instructions both for preparing the body and for actual inoculation were printed in newspapers, as pamphlets, and in at least one almanac, all directed at the literate but not learned ordinary New Yorker. Socially, then, inoculation was a practice people were free to use or not, depending on their own judgment. Inoculation escaped being swept up into any of the multitude of contentions characteristic of New York, and hence its adoption depended solely on its successful integration into familiar ways of healing.[26]

Indeed, because inoculation became accepted reasonably easily, innovations in it then made it that much more widely accessible. Some people in rural areas as early as 1738 took in inoculated people as boarders, so that they could go through their illness in isolation without fear of spreading it to others. In 1757, for example, the Reverend Samuel Johnson, president of King's College in New York City, sent a boy from his household away to a Mrs. Holland to be inoculated. Some women, apparently, became known as skilled smallpox nurses. A handful of doctors, too, specialized in inoculation. In 1758 a Long Island doctor, George Muirson, shared his treatment methods with New York newspaper readers, saying he had inoculated hundreds without one death. He had, he said, originated his method by using it on himself during the 1731 epidemic. In 1768 Dr. Uriah Rogers began advertising his inoculation hospital in Norwalk, Connecticut, just on New York's border, where he would usher people through the smallpox from preparation until the final scabs fell off, all for four pounds. By 1771 he was advertising that he had inoculated "upwards of 1000" without a single loss. By the 1770s, in fact, people from colonies where inoculation was not yet widely accepted began to travel to New York to have the procedure done.[27]

And because New Yorkers grew so familiar with inoculations, improvements to the procedure developed elsewhere found a ready audience in the colony. In 1770 the *New-York Journal* reported that a Mr. Latham, a surgeon attached to the Eighth Regiment of the Foot, had taken a house on Broad Street in New York City and was prepared to offer the new "Suttonian" method of inoculation. In the 1750s Robert Sutton, a doctor in Essex County, England, modified inoculation by lessening the rigors of the preparation (which made it less time consuming and so more affordable), using his own special formula to purge a person's body in preparation for being inoculated, carefully selecting exudates from mild cases of the disease, and making shallower incisions, which were then less prone to fester on their own. Sutton's six sons spread out across England, using their father's method, which did indeed seem to be as efficacious as other methods while less expensive and with fewer side effects. Mr. Latham evidently had some considerable success with his Suttonian practice in New York. In 1771 he moved up the Hudson to rent several houses in Clavarack, and by 1773 had taken on partners in Orange Town, Albany, Schenectady, Kinderhook, Red Hook, Salisbury (Connecticut), Worcester (Massachusetts), and Canada. Dr. Rogers of Norwalk expressed the hope that people "will not be amused with Suttonian pretences," but evidently he was out of luck. The Suttonian modification of inoculation was well received in New York, for New Yorkers had already come to accept that inoculation could dramatically alter their experience with the dreaded smallpox.[28]

Over time, then, inoculation became a significant way improvement entered the lives of ordinary New Yorkers. Information about it was disseminated through print and circulated orally as people talked of their experiences. As a mode of improvement, inoculation was adopted and adapted by an assortment of the middling sort, being neither imposed by an elite nor arising as a local "folk" practice. As dramatic and as useful as inoculation was, nonetheless it was a singular event that drew attention to improvement only episodically. The fabric of most New Yorkers' everyday lives was raising crops and livestock and producing the varied goods necessary to households. Here, too, improvement made its inroads, although here the changes were accompanied by conflict and dissension.

*F*or colonists, agricultural improvement was simultaneously the routine aim of any farmer who took over an abandoned Indian site or set up on promising creek bottoms and the particular mark of gentlemen. "[W]e are improving this Wilderness," wrote Colden, who owned several thousand acres in the Hudson Valley, to a European correspondent, "and have in some measure in some places given it the appearance of the Cultivated grounds in Europe." Colden's comment would not have been out of place a century earlier, for land under European-style cultivation was always an improvement over the landscape colonists found when they arrived. Still, as the landscape became increasingly so transformed, and as some men rose to real wealth, particular ways of transforming the land became indicators of social status. Travelers, for example, contrasted the "improved" properties of Lewis Morris in Westchester County or the Schuylers near Albany, wealthy New Yorkers who cultivated beautiful ornamental gardens as well as extensive commercial agriculture, with the "careless husbandry" of other rural New Yorkers. The wheat cultivated by modest farmers along the west bank of the Hudson, observed one such traveler with disapproval in 1769, "looks much thrown out and gullied." Moreover, beyond the symmetry of the crops of a wealthy landowner (crops often enough cultivated with the labor of slaves) and the beauty of their ornamental gardens, well-to-do men were sometimes active agricultural experimenters. In New Jersey, Charles Read, a member of the New Jersey assembly and a merchant, experimented with various methods of dunging on his farm, and recorded advice from both neighbors and European husbandry manuals. In the 1760s John Bard, a doctor in New York City who had invested in land up the Hudson, had his son, a student in Scotland, investigate madder (a source of dye) cultivation. Such men saw themselves as distinct from the common run of farmers, separated by their particularly vigorous pursuit of improvement.[29]

Nonetheless, information on such agricultural innovation was increasingly offered to more ordinary people. Both newspapers and almanacs continued to offer the recipe advice—brief directions for dealing with some problem with health, husbandry, or craft production—that Daniel Leeds had offered in almanacs beginning in the 1690s. What was striking about the print produced in colonial New York was the surge in advice on improvements in the 1760s. Into the 1750s, such brief advice was rare in newspapers and appeared with modest frequency in almanacs. But thereafter they not only printed advice more frequently but more extensively, such as the twenty-one-installment essay on silk production in the *New-York Weekly Post-Boy* in 1767 or the six-page essay on hemp in Thomas More's 1762 almanac. Usually the sources of this advice went unnoted, but sometimes a source like the *Gentleman's Magazine* of London or a newspaper from South Carolina or Ireland was mentioned, testimony to the circulation of this sort of improvement advice in the wider Atlantic world. Books on husbandry by European authors like Duhamel, Lisle, Miller, and Tull were advertised regularly by booksellers like Garret Noel. *The American Instructor, or Young Man's Companion*, a self-help book for would-be farmers, and its variants were advertised frequently by both booksellers and more general shopkeepers. And, especially after the end of the Seven Years' War in 1763 made northern New York safe from hostile French and Indian incursions, new immigrants brought improved crops, livestock breeds, and agricultural techniques into the colony.[30]

Certainly, the commitment to agricultural improvement was uneven. Richard Smith, on his way to survey land he had newly purchased, observed people around Albany using rope made from elm bark, having learned the skill from local Indians, rather than purchasing the hemp rope sold by shopkeepers. Farmers up the Hudson continued to let their cows forage freely, and farmers on eastern Long Island still set fires to burn out the underbrush in woods. The farming practices of communities of Dutch New Yorkers in northern New Jersey and west of Albany apparently continued little changed from the seventeenth century. Some people, then, were content with familiar knowledge and indifferent to talk of innovation. Indians in and around New York continued to adopt crops and livestock with origins in Europe, but when Smith encountered an abandoned Indian orchard, he noted that they had planted the trees not in rows as colonists did but irregularly. Indians had adopted the crop but not the social meaning that orderly rows of trees indicated a proper domestic order. The embrace of agricultural innovation was also tempered by agriculture's perpetual vulnerability to drought and insects. Moreover, people continued to insist that "experience" was the ultimate arbiter of the useful and the reliable. Thomas More hedged his almanac advice in 1750 with the comment, "[W]herever they fail, you will do well to consider, that possibly our Climate may alter the Virtues which were found to be in those Things elsewhere." As with bodies about to be inoculated, places were peculiar, too; nothing worked everywhere the same.[31]

Nonetheless, the surge in printed advice for farmers in the 1760s indicates some rise in demand. Improvement had certainly always been desirable. "[I]t behoves our Young Women to procure Gardens and our Young Men Orchards," advised Daniel Leeds in his 1712 almanac, "especially since 'tis for the publick good and Improvement of our Country, as well as particular Interest." By the 1760s, newspapers made New Yorkers aware that people in other places were making improvements considerably beyond the ordinary gardens and orchards. Items reported silkworm and indigo experiments in Georgia, woolen weaving in Virginia, linen manufacturing in Boston, burnet and lucern cultivation in Scotland, and hemp cultivation in Britain. When in 1773 Governor William Tryon asserted that it was "good policy to embrace every favorable opportunity of promoting the improvement and cultivation of the Country," what constituted "improvement" had notably expanded.[32]

$\mathcal{A}$s the pronouncement of the governor indicates, to some degree improvement was a deliberate policy of government officials. At the level of the town and at the level of the colony, officials made possible the building of mills, wharfs, and shipyards that sustained economic expansion. And granting water rights or public land to private persons necessarily gave opportunities to some while barring them for others. Rules for weights and measures, for ensuring the quality of exports, for the licensing of taverns, and the situating of tanneries constrained individuals in their pursuit of profit. And always, access to land was access to opportunity, and land grants were in the hands of the governor. Beyond this, laws passed by the General Assembly were subject to the consent of the king. And, as with all other colonies, trade beyond New York's border was subject to the Navigation Acts passed by Parliament. While New Yorkers did not contest the general principle that government provided the framework within which personal economic improvement would go forward, certainly they continually and chronically contested particular rules and outcomes. And, as much as any colonists anywhere, often enough New Yorkers ignored and transgressed regulations that seemed to constrain them unduly, just as many public officials subverted them by pursuing their own and their supporters' interests at the expense of the general good.[33]

Still, to some modest degree, encouraging the independent embrace of improvement became a formal policy of government. During a period of economic stagnation in the 1730s, some New Yorkers, for example, proposed that the government offer bounties to people who adopted hemp cultivation or developed iron mines, as the governments of Massachusetts, Pennsylvania, and New Jersey had done. The current governor, William Cosby, however, was uninterested, as his attention was absorbed in suing a local printer for libel and accumulating land for himself, so this proposal came to naught. In the wake of the dramatic depression that followed the end of the Seven Years War, though, New York's Assembly bestirred itself and sponsored lotteries to raise money to reward hemp cultivation in 1764 and 1765. In 1757 Parliament began offering a bounty to colonists for the production of potash and pearl ash, both of which are made from sawdust and had a wide range of uses from manufacturing glass to bleaching linen. The steady and dramatic expansion of production of all sorts of goods in Britain meant a steadily rising demand for potash and pearl ash. Before the bounty, New Yorkers exported less than forty pounds a year of these two materials; by 1771 they were exporting more than two million pounds a year. Nonetheless, the moderate interest in New York among the powerful in London, compared with more dramatically wealth-producing colonies in the West Indies and the South, and the general orientation of many of New York's royally appointed governors towards the improvement of their own fortunes meant that only a modest amount of New Yorkers' adoption of innovations can be laid to the encouragement of government.[34]

*I*f government actions offered some stimulation to improvement, however modest, in the middle third of the eighteenth century, another social institution emerged in the Atlantic world with the goal of improving society. These voluntary associations, whose objective was the dissemination of useful knowledge, were committed to economic expansion through promoting private production. One such association, the Society for the Encouraging of Arts, Manufacturing, and Commerce, was established in London in 1754. Among its activities, the society offered cash prizes to encourage the production of desirable commodities like silk and wine in the colonies, prizes duly reported by New York's newspapers. In response to the society's offer, James Alexander's son William, who inherited his father's extensive lands in New Jersey, experimented with wine grapes, as did two more modest farmers on eastern Long Island.[35]

Then in 1764, in the midst of the protracted depression induced by the end of military spending on the Seven Years War, a group of New Yorkers established an improvement society of their own. "While Europe enjoys the Advantage of those vast Improvements which Experience has added to the Art of Husbandry," the founding statement proclaimed, "there is scarce an Instance where we have ventured to deviate from the common Road." Responding to "the present deplorable State of our Trade," the first act of the new Society for the Promotion of Arts, Agriculture, and Oeconomy was to offer prizes to spur increases in the quantity, quality, and variety of New York's products. Prizes were offered for flax, hemp, drained swamps, apple seedlings, cheese, mule foals, hops, new meadows, barley, and newly planted hedges, as well as sturgeon and whale oil. The Society also offered to reward expanded craft production with prizes for potash, linen cloth, linen yarn, stockings, hides, shoes, deerskins, tiles, and slate. Describing New Yorkers as hitherto "like Snails, [who] coil ourselves up in the narrow Circle of private Lucre" while other colonists outstripped them, the Society called for a pursuit of private gain that would serve the public by the adoption of all sorts of improvements to agriculture and craft production.[36]

While the announcement of the New York society's formation invited "every real Friend, and lover of his Country" to become a member, and newspaper coverage of one meeting praised the attendance of people of "various Ranks and Degrees, all conspiring to promote the public Interest," the Society was dominated by the wealthy and the powerful. Thirty-seven men sat on the committees of the Society that awarded its prizes, conducted its correspondence, and decided the Society's actions. All but five are easily identifiable as members of the richest landowning families, wealthy merchants, prominent lawyers, and holders of political office, although a handful were less prominent merchants. They laid claim to the status of "gentlemen," men whose entitlement to social and political authority stemmed from their ability to consider the public good above their own private gain, for they were themselves materially secure and had the leisure to consider problems other than their own. In the early modern period the civility that supposedly characterized gentlemen—the willingness to consider problems dispassionately, the honorableness of their testimony, the impartiality of their judgment—had increasingly become the standard by which matters of fact were decided. Members of the New York society asserted that they were just such "public-spirited Gentlemen," those who would extend benevolence towards lesser members of society. They would, for example, make the "Treatises abounding in the most profitable Discoveries" available to local farmers, the sort of people whom they characterized as currently ignorant of such knowledge.[37]

Indeed, among the Society's actions was the reprinting of a pamphlet on hemp so it could be readily available to local farmers. Similarly, they urged that "the most considerable Persons" assist their neighbors' improvement by demonstrating innovations on their own acres. Members also addressed the problem of the burgeoning number of poor in New York City. In 1765 the Society purchased spinning wheels and reels, which would both offer employment to poor women and children and expand cloth production in the colony, and opened a spinning school. That fall it organized a market to sell such locally produced yarn and cloth. And by and large, its prizes were awarded to modest producers. While the wealthy Albany landowner Philip Schuyler won a gold medal for erecting the first flax mill in the colony, a project he almost certainly would have pursued without the encouragement of the Society, Thomas Young, a substantial but not wealthy Long Island farmer, earned a prize for the most apple trees planted. New York's improvement society, then, was intended to be an association in which the greater tended to the lesser in the name of their collective public good.[38]

*H*owever much the rhetoric of the Society proclaimed a straightforward commitment to gentlemen leading the commonalty toward the expanded adoption of improvements for the betterment of all, wider circumstances in the 1760s made for considerable complications. For one, conflict with Britain was rising. When Parliament considered new duties on imported goods and tightening enforcement of duties on West Indian molasses (an ingredient critical to New York rum distillers), New York's remonstrance against the proposed Sugar Act was so vehement that the member of Parliament who had been asked to read it before Parliament in February 1764 refused to do so. As of September 1, New York was forbidden to issue its own currency, and by November, when the Society was formally established, New Yorkers had heard about a plan for a stamp tax, a tax on all printed paper from newspapers to almanacs to legal forms to decks of cards. A year after the Society set up its first market for cloth and yarn, its members joined other New Yorkers in protesting the Stamp Act with a boycott of British goods. During the winter of 1765–1766, trade in and out of the port of New York virtually stopped, unemployment soared, and jails filled with debtors. Then the next summer tenants in the Hudson Valley rioted against their wealthy landlords, with one farmer describing the Livingstons, several of whom were members of the Society, as "the robbers and murderers of common poor people." Far from bowing to the political leadership of the great landowners, the militia in the county with the protesting tenants proved unreliable.[39]

Members of the Society, then, whipsawed between objections to British policies that impeded trade and appreciation for the kind of British military strength that could keep down the rabble. The boycott was lifted in 1766 after the repeal of the Stamp Act but reinstated in August of 1768 in response to the Townshend Acts, another series of taxes on imported goods. Members of the Society found themselves, therefore, consistently in a delicate dance with their correspondents in the London society. On the one hand, they sought to portray New York's society as one that would enhance New York as a trading partner yet not threaten Britain's prosperity. On the other hand, they pointed out the folly of British policies that, they wrote, would only make New Yorkers poorer. Performing the kind of selfless benevolence that would earn the gratitude of the lesser sort and improve everyone's fortunes, while reassuring Parliament and British merchants that New Yorkers' improvement was in their interest too, proved no easy task.[40]

At the same time, while the boycotts of 1765–1766 and 1768–1770 promised to encourage the expansion of production in New York, they also raised disquieting questions about loyalty and virtue. As in every colony along the Atlantic seaboard, the Stamp Act and Townshend Act boycotts in New York elicited an outpouring of sentiment in support of local products. Readers wrote to newspapers promoting tea brewed from the leaves of a local shrub instead of "the poisonous Bohea," and proclaimed the virtue of substituting rye for coffee. A letter writer in 1765 praised "the patriotic and frugal Spirit that begins to reign" in New York, which he said manifested in, among other things, "Town Gentlemen in Country Manufacture." Such wearing of homespun and abjuring of fashionable imports like tea and coffee became visible expressions of loyalty to one's fellow New Yorkers. They also tapped longstanding Protestant values of frugality and plain living, in contrast to luxury and self-indulgence.[41]

The embrace of home production by women, in particular, was emblematic of an upsurge in virtue. New York newspapers regularly reported on spinning parties or praised women like a forty-three-year-old Connecticut woman with twenty-two living children and five grandchildren who spun one hundred twenty-six skeins of wool in a single day, all while nursing her youngest child. This praise was in marked contrast to the castigation of fashionable women, those who, in the words of the almanac maker John Nathan Hutchins, "go naked breasted, and broad stern'd." "Would the Sex in general apply their Hands to the Distaff, instead of the idle Apparatus of the Tea Table," another letter writer piously intoned, "perhaps we need not always be beholden to Asia for our Food, nor Europe for our clothing."[42]

But in the commercial explosion of the eighteenth century, no colonists had outdone wealthy New Yorkers in their embrace of London fashion and conspicuous consumption. The call to embrace coarser local goods was at odds with the visible code by which prominent New Yorkers displayed their status. The call to embrace local productions by high and low undercut the claims of members of the Society to leadership in the promotion of the very activities around which they organized their association.

Neither disputes between New York and Britain nor tensions around consumption led to the demise of New York's improvement society, however. Rather, conflict erupted among the members themselves. The factionalism that characterized New York in the seventeenth century was only deepened and extended in the eighteenth. By the early eighteenth century landowning families with large estates like the Livingstons and the

Schuylers, joined by merchants like the Waltons and DeLanceys, vied for the perquisites of office and economic advantage. These conflicts had no particular ideological content; rather, rhetoric and participants shifted in response to shifting opportunities.

The establishment of the Society for the Promotion of Arts, Agriculture, and Oeconomy, with its proclaimed unity on behalf of the public good, had only been possible in 1764 because at that moment all political parties were united in opposition to the current acting governor, Cadwallader Colden. Colden, a graduate of the University of Edinburgh but a young man of no personal fortune or connection, had been invited to New York by Robert Hunter, the governor who had watched Morgan maneuver his boat, to take up the position of surveyor general of the colony. In 1720 Colden had been appointed to the Governor's Council, from which he eventually ascended to the lieutenant governor's position in 1760, partly through sheer longevity. Consequently, in 1764, while New Yorkers awaited a new appointee, Colden served as the acting governor. And Colden was a man with a strikingly poor political ear. He was particularly loathed for his efforts as the colony's surveyor general to ensure that land grants were done within the law. In the early 1760s he had obstructed the efforts of some New Yorkers to have dubious land titles allegedly acquired from the Iroquois acknowledged as legitimate. In 1764 he had alienated all sides by voiding a jury decision. The next year he went on to embrace the Stamp Act and order it enforced. For this he was burned in effigy. Colden, then, for a brief period provided a focus around which long-standing enemies could unite.[43]

In November 1765 the actual governor, Henry Moore, arrived, and thereafter normal political conflict within New York resumed. Even in 1766, with the Society at its most vigorous, member James DeLancey was negotiating with the Sons of Liberty (one of whose leaders, John Lamb, was also a member) for its support in the next Assembly election. The 1768 elections pitted member John Morin Scott, who was supported by his allies William Livingston and William Smith Jr. (also members) against James DeLancey, John Cruger, Jacob Walton, and James Jauncey (all members too). William Livingston, a Presbyterian, also began to campaign against a proposal to establish an Anglican bishop in the American colonies, a campaign that maligned his rival DeLancey, an Anglican.[44]

If the 1768 elections were vituperative, the 1769 elections were even more so. By this point there was virtually no report of Society activity in the newspapers. After one last call to a meeting in 1770, the Society died away. It had imploded from within, killed by political rivalry. In New York generally, political battles, the suppression of tenants, rifts over sustaining the boycotts, struggles over land titles, and religious disputes made for a bitter colony and one in which the claims of the wealthy to leadership in any realm was increasingly subject to scorn. By 1770 the political and social elite of the colony had abandoned any display of homespun; instead, in Edward Countryman's words, they resumed their "glittering pattern of conspicuous consumption." That year "A.B." published a broadsheet, "Proposals for Erecting and Encouraging a New Manufactory." In it he mocked the pretensions of the wealthy to contribute to the public good with their projects of improvement: his "manufactory" was a tannery where the rich could turn the skins of working people into hides and reap the profits.[45]

Despite the efforts of some New Yorkers to associate "improvement" with "gentleman," and so enhance their claims to leadership in the colony, ultimately their efforts failed. Even if travelers compared the innovative and orderly estates of large landowners with less elegant farming of the more modest and read a social hierarchy of gentlemen and the common sort, few more ordinary New Yorkers seemed to care. The middling sort did not mistake the professions of the members of the Society that they would lead New Yorkers into a world of improvements as expressions of some selfless devotion to the public good. Such professions did not mask the rivalries of wealthier men, nor did they persuade more modest New Yorkers to look to them to tell them what improvement was.

Nonetheless, by the 1770s improvement had clearly come to New York. Men of means, like Philip Schuyler, invested in projects like flax fields and flax mills and planned others, like a canal to link Lake George and Lake Champlain. Middling people also expanded what they grew and made, as the statistics on hemp exports and the ready claimants for the Society's prizes show. In 1767, the *New-York Weekly Post-Boy* reported on the spectacular output of linen by Ebenezer Hurd, "who has rode Post for 40 years." Newspapers and almanacs carried news of and advice on various improvements, and booksellers regularly stocked books on husbandry and navigation. In 1770 a more modest group of men established another improvement society, with more modest goals. The Marine Society of New York would sponsor better maritime maps, hold discussions of better navigational methods with its sea captain and navigator members, and give charitable support to the distressed among seagoing men and their families. As Elizabeth Fenn points out, soldiers from the Middle Colonies would be among the least vulnerable to smallpox when the Revolution broke out; Dr. Muirson, Mr. Latham, and their colleagues had made inoculation accessible throughout the colony.

Some claimed that even the climate of New York was improved. "[W]hoever compares the present state of the air, with what it was formerly, before the country was opened, cleared and drained," wrote one man, would find that it was, in every way, improving.[46]

Still, the increased vigor of improving practices did not utterly transform the relationship of New Yorkers to the natural world. For one, the natural world itself still wrecked ships with hurricanes, farmers with drought, and people everywhere with illness. Efforts to innovate sometimes failed and so reinforced skepticism toward the untried. Experience was still the measure of the useful, and what worked for one might not work for another. If in general life was easier and safer in 1770 than it had been a century before, there were plenty of experiences that countered too complaisant a faith in improvement. Moreover, few, if any, New Yorkers doubted that the hand of God lay behind everything. "Extreemly hot and dry," wrote Mary Cooper, a Long Island farmwoman, in her journal on July 8, 1769. Echoing Alida Livingston, she added, "If the Lord dose not look in mercy upon the earth and send some rain we shall soon perish." Two years later, her adult daughter returned safely from undergoing smallpox inoculation, but here, too, it was God as much as mankind who was responsible. "The Lord has brought my daughter home to me, well of the small pox," Cooper wrote. "What shall I render to the Lord for all his mercys?" And even as inoculation became ordinary, Cooper herself was unsure how far to trust it. Cooper was terrified of contagion for a week after her daughter's return, even though the daughter had passed through the illness completely.[47]

Overall, then, an increased confidence was evident among many New Yorkers by the 1770s that through their own efforts they could master enough of the natural world to improve their personal circumstances. Yet this was tempered by considerable wariness towards the untried, by the belief that ultimately one's fate was in the hands of Providence, and by skepticism towards those who made improvement a badge of social superiority and entitlement to govern.

CHAPTER 5

Refinement

The woods, the fields, where'er I walk,
In solemn truths declare,
The blushing rose upon its stalk,
Say, there's no resting here.
Let heaven-born souls then look above,
Where joys eternal spring;
Let them adore with fervent love,
Their Maker, God, and King.
Roger More, Poor Roger.
The American Country Almanack for . . . 1770

"[*B*]ut this I must beg as a favour of you," Jane Colden, a woman living in the Hudson Valley, wrote to Charles Alston, a professor of botany in Scotland, "that you will not make any thing publick from me till (at least) I have gain'd more knowledge of Plants." Colden was a collector and describer of plants, one of the first women anywhere in the Atlantic world to master the new Linnaean method of plant classification. The Linnaean method allowed for discrimination between plants that looked superficially similar, and for clustering plants that looked superficially different. It thus revealed organizational patterns in the world's flora that escaped the notice of people unfamiliar with the fine details of plant anatomy. Colden was among those few who examined plants, and indeed recorded her observations in a notebook, carefully detailing each part of a plant. In the 1750s she described 352 New York plants, including two previously undescribed by botanists. Moreover, her botany dramatically expanded her circle of acquaintances, both people she met face to face when they visited the farm where she lived and people, like Professor Alston, whom she knew only through correspondence. For Jane Colden, botany was both a new way of knowing the natural world and an enhanced mode of sociability.[1]

In the eighteenth century, ways of knowing the natural world that we now associate with the early development of formal science became more prominent in the Atlantic world. As with botany, new methods of organizing information enabled the development of encyclopedias of natural phenomena. Encyclopedias of plants, for example, aimed to be compendiums of all the world's species, laid out without regard to the particularities of place or use, but solely with regard to making comparisons among them. New ways of reproducing natural phenomena, as with electrical experiments that developed in the 1730s, gave rise to new explanations that seemed to open up phenomena like lightning to human mastery unlike anything that had come before. This invigorated attention to categorization and experimentation took medicine off in new directions too, with doctors with formal MD degrees increasingly claiming knowledge distinct from that of common doctors. In New York, as elsewhere, these new approaches to the natural world became increasingly visible.

These new approaches to the natural world were embedded within new forms of sociability characterized by "refinement." Refinement entailed making distinctions between the elevated and the crude, drawing particular sorts of boundaries between "us" and "them." As such it was increasingly a marker of the "gentle" class, claimed by ladies and gentlemen to distinguish themselves from the vulgar. All around the rim of the British Atlantic, people with claims to gentility increasingly substituted porcelain for pottery, chairs for stools, and elegant fashions for plain homespun. Homes were divided into rooms for elegant hospitality and rooms for personal privacy, hung with wallpaper and portraits, and set off with ornamental gardens. Ladies and gentlemen read poetry and novels along with their Bibles and their almanacs. They drank tea, made genteel conversation, learned fashionable dances, and listened to performances of Handel. If improvement had been relatively unsuccessful at marking off gentlemen in New York, refinement intended to separate more distinctly people of refined sensibilities from those of the rude.[2]

The surge in such refined consumption and sociability in the eighteenth century, in New York as elsewhere, was striking and has called for explanation. Certainly, the sheer pleasure of pretty fabrics and engrossing novels explains some of it, and the rising volume of commerce explains how people paid for these things. But it is also clear that the demand for these goods preceded people's ability to pay for them, and hence the rising volume of commerce is as much an outcome as a cause of the surge in refined consumption. Historians have suggested that displaying such gentility was functional as well as pleasurable. For one, it facilitated social relations in a mobile world in which people as well as goods circulated. Shared conventions around a tea table and familiarity with the same poets and essayists made common ground among strangers. For another, it was a sign of creditworthiness in a world in which banks and insurance were barely developed. One displayed one's own prosperity, and so hopefully one's worthiness to be extended further credit, through the display of worldly goods. Through being fashionably au courant, one showed one's worldly savvy. Further, refinement was a sign of a cultivated soul, an elevated state that was "honor" in a man and "virtue" in a woman. Such men and women were worthy of trust because they were people of reliable character.[3]

Beliefs and practices concerning the natural world in the eighteenth century incorporated some of the elements of refinement. Increasingly, people of refined tastes collected exotica that they showed off in cabinets of curiosities, because such cabinets displayed their familiarity with far-flung places and their ability to contemplate strangeness with equanimity. While such exotica had long been collected by the cognoscenti, wider circles of people did so in the eighteenth century. They increasingly read books about natural history and collected engravings of natural phenomena. Gardens too became sites for growing plants from the four corners of the globe. The plethora of natural history specimens, plant and animal, generated more ardent efforts to categorize them, to discern their natural patterns. In turn, the contemplation of these patterns, the natural laws by which all living things were related, were intended to enhance one's reverence for God, the author of this book of nature. In contrast to the religious "enthusiasm" of the vulgar or to their uncomprehending astonishment before natural oddities, people of refined sensibilities were moved to calm contemplation and serene worship by the wonders of the natural world. And the accompanying sociability, the shared conversations about such natural phenomena, were intended to serve as a template for social relations among people of the gentle classes generally, in contrast to the competitive, contentious realms of commerce and politics. Although actual relationships among collectors, categorizers, and cataloguers was often as competitive as they were in other realms, nonetheless the ideal was an allegiance to a dispassionate and selfless pursuit of the true nature of the natural world.[4]

In New York, the adoption of such refined sensibilities toward the natural world was apparent but relatively modest, for refinement in and of itself never became central to demonstrating one's creditworthiness. A handful of New Yorkers, like Jane Colden, engaged in the newly systematized botany of Carl Linnaeus, and others apparently purchased some of the elegant natural history books offered for sale by New York City booksellers. Still more attended shows of various curiosities and the increasingly popular electrical demonstrations, although it is hard to know how many contemplated these with a refined sensibility. The most vehement claim to an elevated understanding of nature, though, came from an influx of doctors with formal MD degrees, men who claimed the status of gentlemen because of their erudite understanding of the human body and disease. New Yorkers were generally skeptical of these claims, understanding the claim to being a gentleman as merely a claim to higher fees. Nonetheless, by the 1770s some of these gentlemanly doctors had managed to establish themselves with a decent degree of economic security, for to some degree New Yorkers' understanding of the body had shifted in ways that made their claims plausible. If few New Yorkers made any sustained effort to investigate the natural world, the decorum of natural history, as well as the larger claim that a nature that yielded up its natural laws to scrutiny lifted one's eyes to God, had become, at the very least, familiar.

$\mathcal{A}$s a realm of knowledge of the natural world, botany exemplified how refinement came to characterize aspects of science in the eighteenth century and how, in turn, such aspects of science became a mode through which people could express a refined sensibility. Plants displayed a pleasing variety, their subtle details rewarded a discerning eye, and their beauty aroused reverence for their creator. Collecting and displaying exotic plants, either in herbaria or, better still, in elegant gardens stood as a witness to a process by which Europeans were drawing all the world into relations of commerce, alliance, and subordination. The rising number of useful foreign plants—tobacco, sugar, maize, tomatoes, peppers, cinchona—affirmed that Europeans were indeed living in an age of progress. If in previous centuries interest in botany had been the province of a handful of doctors and university scholars, by the early eighteenth century there was a rising interest in botany generally. Collectors in England, men who often belonged to the Royal Society, sponsored trips to Chesapeake Bay in the 1690s and to the southern colonies in the 1710s. A handful of women, too, participated actively as collectors or patrons. Increasingly, elegant people perused natural history books, exchanged plants as gifts, rambled their local countryside cataloguing their own local plants, and grew exotic plants in their gardens.[5]

Very little of this heightened interest in botany was evident in New York. If early colonists like Adriaen van der Donck and Daniel Denton had touted the promise of plant life in the colony, believing that some surely would be useful, actually discerning the use of particular indigenous plants was very slow. By the middle of the eighteenth century, a few more indigenous plants had come to be used by colonists. The root of one low-lying plant in the Catskills was used to treat colic, and the leaves and berries of a shrub were made into a plaster for cancers. Men in search of promising land routinely assessed variations in types of trees, believing that different species of tree were guides to the type of underlying soil, as Richard Smith did when he surveyed land he had bought near the Mohawks in 1769. But if New Yorkers had slowly grown more familiar with at least some local wild plants, most were indifferent to systematizing what knowledge they now had.[6]

In the 1720s a New Yorker with a different sensibility dreamed of writing a natural history of the colony. Cadwallader Colden, the man whose unpopularity would so unify political factions in the 1760s, had arrived in the colony in 1719 with a degree from the University of Edinburgh. Having failed to establish himself as a doctor in London and as a merchant in Philadelphia, Colden had been fortunate in catching the eye of Governor Hunter, the man who had watched Morgan row his boat off Manhattan. Hunter appointed Colden New York's surveyor general in 1720, and soon after Colden was appointed to the Governor's Council. In his first decade in the colony, Colden had found in Hunter and in his successor, William Burnet, also a Fellow of the Royal Society, men with kindred interests in the natural world. Under Burnet's sponsorship, Colden had written reports on New York's climate and on its economy for the Lords of Trade in London, published a history of New York's relations with the Iroquois, and joined Burnet and James Alexander, another Scot, in determining the longitude of New York City, the report of which was gratifyingly published in the Royal Society's *Philosophical Transactions.* However, despite having studied botany at the university, Colden later wrote, "I found so much difficulty applying it to the many unknown plants that I met with everywhere that I was quite discouraged and laid aside all attempts in that way." Like most colonists, Colden looked around him and saw in the welter of local plants only an undifferentiated wilderness.[7]

It was indeed genuinely difficult to make sense of this abundance of unfamiliar plants, for by this point catalogues of plants ran into many volumes and were generally organized alphabetically. It was impossible to look up an unfamiliar plant if you didn't already know what it was called, and daunting to compare it with what was by now thousands of plant descriptions. Moreover, the names themselves were not stable. In 1748 Naphtali Frank, a young New Yorker from a merchant family doing business in London, asked his mother to send him cranberry plants and the seedlings of several indigenous trees. He evidently intended to use them as gifts for his business contacts. His mother dispatched what he asked for as best she could, but complained "I beg You will Explain Your Self when You Send for anny of these things for I am neither Gardener nor botanist and Consequently Understand Very Little of the Matter." Abigail Frank's problem with identifying plants did not stem merely from her own indifference to botany. Even among knowledgeable collectors and nurserymen in Europe, confusion over names was common. Sometimes a single plant had been named several times over as it entered Europe at different times and in different places, or different plants were given the same name if the written description of one seemed like another. One of the tenets of botany was that the plant world revealed the harmony and order of God's universe, but arranging plants in some sort of intelligible order was in fact very hard to do.[8]

In 1735 Carl Linnaeus, a Swedish naturalist working in the Netherlands, where interest in natural history was particularly strong, devised an ingenious solution to this problem of botanical cataloguing. A man of prodigious memory, Linnaeus recognized that each species of plant had a stable number of male and female reproductive parts, its stamens and pistils. Species with the same number of male parts could all be grouped together in a single class. Within that class, species with the same number of female parts could be clustered in the same order. Then within each order, plants could be divided into genera by characteristics of stems and leaves, and then distinguished further into individual species by other minor but consistent characteristics. Although some plants, like ferns and mosses, did not lend themselves easily to this system, the vast majority of the world's plants did. An unfamiliar plant could be identified (or recognized as previously unknown to Europeans) by counting its stamens and pistils and then comparing it solely with the plants grouped within the same class and order. Linnaeus's system was adopted enthusiastically in the Netherlands, and, although it would be another twenty years before Britons would so unequivocally embrace Linnaeus, his system circulated in New York almost immediately. While New York had long been politically part of the British Empire, Dutch New Yorkers sustained ties with the Netherlands. Soon after Linnaeus developed his system, Isaac DuBois, the son of the Dutch Reformed minister in New York City, arrived at the University of Leyden to go to medical school. In 1740 he returned home with his medical degree, and with him he brought Linnaeus's books.[9]

Sometime in the winter of 1741/42 DuBois showed Linnaeus's books to Colden. Colden now lived on a farm in the Hudson Valley, in relative isolation. His intellectual kinship with New York's governors in the 1720s

had extended to ardent support of their politics and consequently he had
made many enemies. In 1728 a new governor replaced Burnet, and Colden's
enemies got the new governor's ear. While Colden lost neither of his formal
positions, he was displaced from circles of power. The next year he and his
wife Alice moved to land he had bought about ninety miles north of New
York City. All through the 1730s they developed their farm and raised their
eight children, with Colden's intellectual ambitions seemingly dead. But
when DuBois showed him Linnaeus's system, Colden recognized how it
would transform the wild plants around him into legible objects. During
the summer of 1742 Colden examined plants that grew on or near his farm,
counting their stamens and pistils, and writing up complete descriptions
of their parts in Linnaean sequence for ninety-one of them. Although he
could not identify the genus names of a half dozen of these plants, he successfully named 231 local plants, thus bringing them into the realm of what
to learned Europeans signified the known. Linnaeus's system had enabled
Colden to master a part of his surrounding environment he had previously
found unintelligible.[10]

Whatever pleasures Colden's newfound botanizing may have held for
him, its larger significance was that it opened the door to a wider world of
sociability. He had had some minor contact with Peter Collinson, a London
merchant with a passion for botany. Colden's new botanical interests now
led them into genuine friendship. Collinson wrote to his own North American collector, John Bartram, a Pennsylvanian who had made collecting trips
into New York (and who had made consistent complaints about the difficulty of getting New Yorkers' cooperation), about Colden. Bartram then visited
Colden on his next trip into the Hudson Valley. With Bartram, Colden had
a friend in the flesh. DuBois sent Colden's descriptions to the preeminent
naturalist at Leyden, Jan Frederick Gronovius, who was delighted with this
new correspondent from North America. Gronovius sent Colden's descriptions on to Linnaeus himself, who had returned to a university post in Sweden. Linnaeus was as pleased as Gronovius had been. Linnaeus arranged to
have Colden's plant descriptions printed in the *Acta* of the Swedish Royal
Academy, began a correspondence with him, and even named a newly discovered plant genus after him, a mark of real distinction in the botanical
world. After Colden's long intellectual isolation, botany furnished him with
a circle of friends and with esteem, a marked contrast to his experiences in
New York.[11]

Linnaeus also dispatched one of his students to this newly promising
area of the world, for his encyclopedic project of cataloguing the world's

plants required specimens from all over. Peter Kalm arrived in Pennsylvania in 1748 and visited Colden almost immediately, bringing him letters from Linnaeus with follow-up inquiries about Colden's flora and a copy of the book that listed the new genus *Coldenia*. Kalm then went on to Montreal, returned to Pennsylvania, and made a trip to Niagara. As he went, he queried people on their plant knowledge. He collected information on a number of healing plants, some of which, he reported, people had learned from Indians and others from finding indigenous plants similar to European species. In Albany, for example, he visited with the doctors Rosabom, father and son, who told him of several Indian cures but also treasured a sixteenth century Dutch herbal as their guide to healing plants. But even though Kalm found some local informants useful, he commented that New Yorkers generally paid little attention to natural history, "that science being here (as in other parts of the world) looked upon as a mere trifle and the pastime of fools." Colden indeed had been lucky to find fellowship through correspondence, for there was little of it locally.[12]

While Colden had been happy to see Kalm, and very pleased to see his name in print (he proudly inscribed "Ex Dono Autoris" in the book with the *Coldenia* in it), he had had in fact little time to spend with Kalm. In 1746 Colden had been drawn back into New York politics after years on its periphery. The reason was a dispute over the conduct of King George's War. The Crown directed New York's governor, George Clinton, to attack Canada, which directly threatened New York from the north. Albany traders, forbidden by law to trade with the French, nevertheless did so regularly, finding the trade easy and profitable. Other New Yorkers disliked the threat of war taxes and disruption to their farm work occasioned by a militia call. Opposition to Clinton was led by the powerful chief justice of the colony, James DeLancey (the father of the James DeLancey who was a member of the improvement society in chapter 4). Clinton needed a local ally, and he turned to Colden. It was a disastrous mistake. As he had in the 1720s and as he would again in the 1760s, Colden managed to alienate practically everyone. By the end of 1750 he was back on his farm, still in possession of his formal positions but again on the periphery of power.[13]

His botanical circle still valued him, though, for they had no stake in New York's politics. Bartram still visited and Collinson still wrote, eager for anything new. Linnaeus would have welcomed anything Kalm had missed, or details Kalm couldn't have seen. But Colden had nothing to contribute. He was now in his sixties and described himself as an old man. He was too old, he wrote, to go out searching for new plants, his eyes too bad to make out the fine points of a plant's anatomy. For this he needed a proxy, and in 1752 he found one: his fourth child, his twenty-eight-year-old daughter Jane.[14]

In choosing a daughter to do more than serve as a polite audience for such innovations in scientific knowledge, Colden could draw on the unusual but certainly not obscure model of the learned woman. While few women in the middle of the eighteenth century were producers, rather than consumers, of the new sciences, exceptions like Laura Bassi, a professor of natural philosophy in Italy, were renowned enough to be mentioned in colonial newspapers. Beyond the model of the learned woman, Colden drew on a more fundamental understanding of the relationship between gender and learning. In 1742, when he first discovered Linnaeus, Colden suggested that his work be translated from Latin into English to make it available both to less-learned men and to women. "The Ladies are at least as well fitted for this Study as the men," he wrote, "by their natural curiosity & the accuracy & quickness of their Sensations." For early modern Europeans, intellectual differences between the sexes were rooted in the physical body. Women's bodies were relatively cool and moist, and hence women's sensations were, as Colden expressed it, quicker, their minds livelier. Drier, warmer male bodies, by contrast, were firmer and so male minds were more capable of sustained intellectual activity. Such understandings were commonplace, but not therefore inflexible. Particular persons were understood to vary in their degree of moisture and temperature, and education could do much to shape the expression of bodily aptitude.[15]

In any case, the model of the learned woman and assumptions about women's bodily nature allowed Colden to select the person who was, in fact, the best choice. His eldest three children were long since married and another had died. Jane was apparently a capable domestic manager (described as "Carefull" with none of the "Wild Giddy humours" attributed to a younger sister) with "an inclination to reading." And indeed, when she took up botany under her father's tutelage, her skill and engagement soon surpassed his.[16]

Over the next five years, Jane Colden rambled through fields, woods, and bogs near the farm and wrote descriptions of over 350 plants. She used a magnifying glass to see parts too small to see otherwise and dug up roots so she could describe them. She entered her descriptions in a notebook, one plant to a page, laying out each description sequentially by plant part (illustration 5). She counted the stamens (called in English "chives") and the pistils so that every plant could be placed within its proper Linnaean category. Thirty-five of the plants in Jane's flora had been described by her father, but although many of his had too few details to compare them with Jane's, at least 175 were hers alone. And even for plants he had described, she saw details he had missed, like the delicate fibers that accompanied the central rib of a *Polygala* leaf or the rough grains inside the petals of a *Uvularia*. And at least two of the plants she described were original with her, for they had previously gone unnoticed by those collecting North American flora. Jane left little from these years beyond the notebook itself, yet botany seemed to have brought her real pleasure, for it was an absorbing activity, perhaps in contrast to the tedium of farm life, one for which she had a real talent and for which she earned the admiration and gratitude of her father.[17]

Indeed, Jane's work revived her father's contacts with the botanical world and even expanded them. Cadwallader wrote old friends like Peter Collinson immediately and showed off her work to John Bartram when he stopped by on a collecting trip. He sent samples of her work to Gronovius, offering her services as a collector. Collinson brought both Cadwallader and his daughter to the attention of other prominent London naturalists, like the botanist John Ellis and the physician John Fothergill. Cadwallader then sent Fothergill observations on a local fever that Fothergill published in a London medical journal, another boost to Cadwallader's self-esteem. When another of Collinson's colonial contacts, Alexander Garden, a Scots physician re-settled in South Carolina and an ardent naturalist, took a visit to the northern colonies, he paid a visit to the Coldens and was delighted with their company. Garden widened Cadwallader's contacts still further by introducing him to faculty at the University of Edinburgh, where Cadwallader's contacts had long since lapsed.[18]

When Garden visited the Colden farm, he found a small pink-flowered plant in nearby woods. Jane showed him the description of the plant that she had already written up, suggesting that its characteristics made it a previously unknown genus. Standing on the etiquette of discovery, Jane proposed to call it "Gardenia, in compliment to Dr. Garden." Garden sent both her description and his own to Edinburgh, where they were published in a learned journal in 1756. At the time, botanists disagreed that this was a new genus and consequently it was assigned to an existing genus, *Hypericum,* so she did not get the opportunity to name it. (She was correct, but by the time this was recognized in the nineteenth century and reassigned, her role in discovering this plant was long forgotten.) Nonetheless, she was the first woman anywhere to have a Linnaean plant description of a previously unrecognized plant published. She corresponded with Edinburgh's botany professor, Dr. Alston, and joined her father in his further correspondence with Garden. All

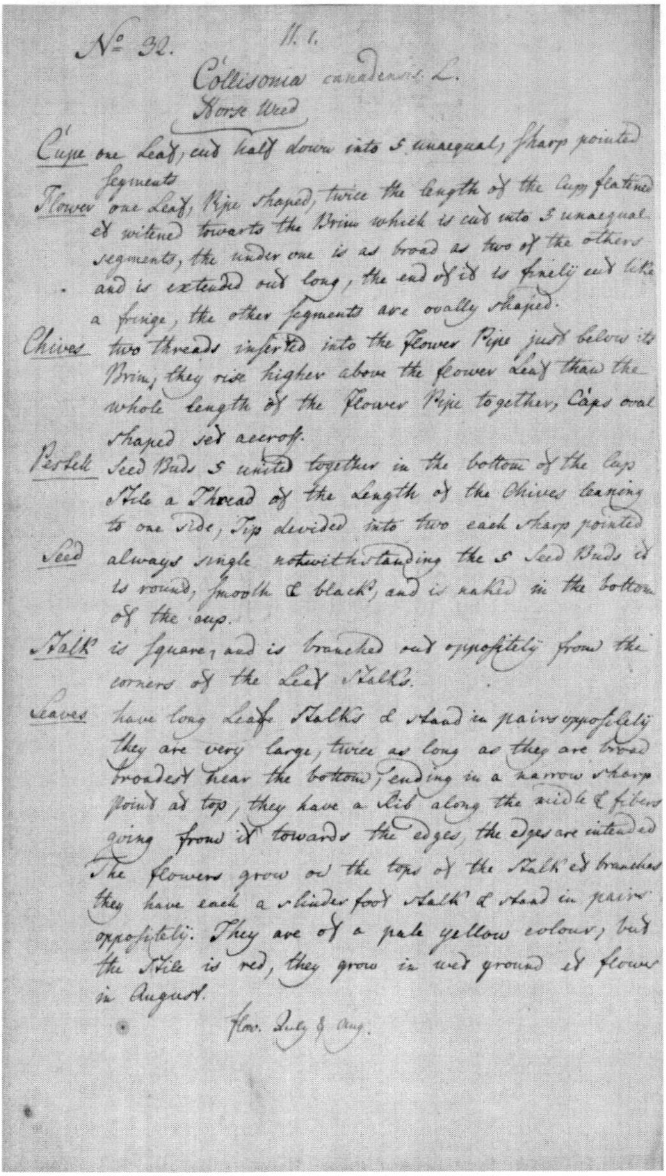

N:º 32.   II. I.

Collisonia canadensis L.
Horse Weed

Cup one Leaf, cut half down into 5 unequal, sharp pointed
    segments

Flower one Leaf, Pipe shaped, twice the length of the Cup, flattned
    & widened towards the Brim which is cut into 3 unequal
    segments, the under one is as broad as two of the others
    and is extended out long, the end of it is finely cut like
    a fringe, the other segments are ovally shaped.

Chives two threads inserted into the flower Pipe just below its
    Brim, they rise higher above the flower Leaf than the
    whole length of the flower Pipe together, Caps oval
    shaped set across.

Pestil Seed Buds 5 united together in the bottom of the Cup
    Stile a Thread of the length of the Chives leaning
    to one side, Tip devided into two each sharp pointed

Seed always single notwithstanding the 5 Seed Buds it
    is round, smooth & black, and is naked in the bottom
    of the cup.

Stalk is square, and is branched out oppositely from the
    corners of the Leaf Stalks.

Leaves have long Leaf Stalks & stand in pairs oppositely
    they are very large, twice as long as they are broad
    broadest near the bottom; ending in a narrow sharp
    point at top, they have a rib along the middle & fibers
    going from it towards the edges, the edges are indented
    The flowers grow on the tops of the Stalk & branches
    they have each a slender foot Stalk & stand in pairs
    oppositely: They are of a pale yellow colour, but
    the Stile is red, they grow in wet ground & flower
    in August.

            Flow. July & Aug.

ILLUSTRATION 5. Jane Colden, No. 32: Collisonia, "Flora Nov. Eboraensis." This page from Colden's notebook shows the care with which she described individual specimens of plants. Such descriptions allowed for comparisons between plants and their classification. (*Reproduced with permission of The Botany Library, Natural History Museum, Great Britain.*)

through the 1750s, then, botany brought the Coldens into a circle of sociability sustained by letters, specimens, and seeds, one in which their mastery of plants, their attentiveness to their natural order, and the decorum of refined attentiveness to the natural world made them welcome.[19]

Then local events overtook them, disrupting their botanizing. In 1757, fearing raids because of the Seven Years War, the Coldens left their farm and moved to Flushing, Long Island, not far from New York City. Jane spent the next year learning Latin in order to read formal botany and drawing illustrations for her notebook. Like her frail younger brother, she also had herself inoculated for smallpox, a procedure that would preserve both her life and her complexion. In 1759 she married. Her husband was a Scots physician, a widower twenty years her senior. Her father assumed that her marriage would end her botanical interests, as if, having left his household, she would cease to do the work of his household. We don't know if that's the way she saw it, for her marriage lasted only a year. She died in March 1760.[20]

At the same time, her father ascended to the position of lieutenant governor, partly because of sheer longevity, but primarily because of his continuous stalwart defense of the king's prerogatives. It's hard to know how much of a role his botany played in this, the degree to which his connections with naturalists like Peter Collinson and John Fothergill enhanced his reputation in London among the politically interested. But certainly botany sustained his own self-esteem, and affirmed his place among gentlemen and thus among those most fit to guide the government.

*T*he Coldens were unusual, for their local circumstances and Cadwallader's inclinations pushed them to embrace a formal science in which few other New Yorkers were interested. Refined interpretations of natural phenomena more commonly came to a wider audience of New Yorkers through another venue, lecture demonstrations of electrical phenomena. In the eighteenth century electricity was a novel subject for investigation. Early in the eighteenth century a handful of men, like Francis Hauksbee in Britain, began to investigate odd phenomena like the sparks generated by friction on certain materials. By the 1740s men with such interests had developed a host of apparatus with which to experiment on the phenomenon of electricity, experiments that were widely reported in London publications like the *Philosophical Transactions* and the *Gentleman's Magazine*.

Some men then began to make a living or at least supplement their living by touring the provincial towns and seaports of the British Atlantic. In 1744 Archibald Spencer, an Edinburgh-trained physician, was the first to offer electrical demonstrations in New York City, followed by Richard Brickell in 1748, the Philadelphians Lewis Evans and Ebenezer Kinnersley in 1751 and 1752 respectively, and an Irish Quaker, William Johnson, in the 1760s. Advertised to "Ladies and Gentlemen" as "a rational and agreeable Entertainment," such shows promised to demonstrate that electrical phenomena followed orderly natural laws. "[T]he wisdom of Providence," a Rhode Islander wrote in encouraging New Yorkers to attend Kinnersley's lectures, had now made intelligible a natural phenomenon "which has been a Mystery, wrapp'd up in clouds and thick darkness." Electrical shows revealed this previously mysterious phenomenon in all its clarity and regularity.[21]

If Colden had used botany to demonstrate his gentlemanly refinement, New Yorkers well recognized that Benjamin Franklin from nearby Pennsylvania had been visibly successful at socially elevating himself via electricity. In 1746 Franklin had received a glass tube designed for electrical experiments from Peter Collinson. For the next few years Franklin and his circle experimented and innovated, with Franklin developing the key conception of positive and negative charges. Franklin's ideas about the nature of electricity and his accounts of the group's experiments were, in turn, conveyed to Collinson, who saw that they were published in the *Philosophical Transactions*, in the *Gentleman's Magazine*, as a series of pamphlets, and finally collected in a book. Franklin was not the first to suggest that lightning was an electrical phenomenon, but his reasoning and his proposal for testing his reasoning were widely circulated. When his experiment was done in France in 1752 and was successful, the resulting acclaim was reported to New Yorkers, as were his honorary degree from Harvard and the Copley medal he received from the Royal Society. Franklin's investigations were also the basis for a striking new improvement, the lightning rod. New York newspapers reported glowingly on the utility of lightning rods in deflecting the damage done by lightning. In 1754 the *New-York Weekly Post-Boy* published a poem in honor of Franklin written by "C.W." of Cooper-River, South Carolina. While lightning, the poet wrote, used to terrify him, now he could contemplate it with composure.

No fire I fear my dwelling shou'd invade,
No bolt transfix me, in the dreadful shade.
No falling steeple trembles from on high,
No shivered organs now in fragments fly.
The guardian point erected high in air,
Nature disarms, and teaches storms to spare.

In pointed contrast to the reactions of his slaves and of animals that he also described, C.W. could display the sensibility suitable to gentlemen when faced with this natural phenomenon.[22]

By the end of the 1750s, Franklin's work with electricity had publicly transformed him into the model of a gentleman of refined sensibility: his sagacity had penetrated to the order underlying seemingly capricious natural phenomena; his communications on electricity had rendered it intelligible to the literate world; and his lightning rod had contributed concretely to the public good.

Electrical shows were intended to allow the audience to cultivate this refined sensibility by observing the demonstrator's control over electrical phenomena. By the 1740s demonstrators used glass balls with a leather or cloth piece pressed to them that could be rapidly rotated to throw off sparks, and foil-covered, water-filled jars that could accumulate electricity and so discharge powerful jolts. Demonstrators could then make an artificial spider dance, a toy cannon fire, and a gilt crown on a picture of King George glow. They could make "lightning" emerge from a paper cloud and strike a wooden model of a church, which would gratifyingly split apart, and then show how the church was spared by a lightning rod on its steeple. Demonstrators hoped audiences would draw the right conclusions. "[T]he knowledge of Nature tends to enlarge the human Mind," wrote the demonstrator William Johnson in one of his advertisements, "and give us more noble and exalted Ideas of the God of Nature."[23]

*O*f the New Yorkers who attended electrical shows it is difficult to know how many left the shows with "more noble and exalted Ideas of the God of Nature." Amid the dancing spiders and firing cannons was a demonstration called the "Salute Repuls'd," in which a lady's cheeks and lips were charged so that no man could kiss her, no matter how hard he tried. While perhaps audiences were moved to the refined contemplation of nature through watching the "Salute Repuls'd," electrical shows were undoubtedly as much like familiar "wonder" shows as they were demonstrations of God's orderly universe. In New York as elsewhere, shows offering oddities of nature continued to have considerable appeal, the more shocking the better. Some of these were of local origin, like the seven-legged calf to be seen at the house of George Hopkins in 1752. Others were part of the rising pace and extent of commerce in the wider Atlantic world, such as the tightrope-walking monkey put on display by Edward Willet in 1751. Mechanical curiosities too, like one depicting landscapes and seascapes ("nearly imitating Nature") also were periodically the basis for shows for the curious. John Bonin, for example, gave shows for four shillings featuring an optical machine that "represents Perspective Views of most of the famous Palaces and Gardens of England, France, and Italy," clearly places of refinement. However elevated Bonin's subject, and despite his lowering the price to two shillings and touring Long Island, he evidently had trouble exciting enough curiosity to stay in business. In May 1750 Bonin announced that he was opening a shop, assuring his creditors that he would pay what he owed. Whatever the hopes of exhibitors, once their curiosity was satisfied, New Yorkers evidently felt little need to view such oddities repeatedly.[24]

In 1751 Bonin was back in show business with another curiosity, a live porcupine. From the mid-eighteenth century on, one of the subtle alterations in the characteristics of wonder shows in New York was the appearance of North American wild animals. In 1754 people in New York City could see an alligator, in 1756 the skin of a twenty-one-foot snake killed in the Allegheny Mountains, in 1757 a buffalo, in 1761 a "tiger," also from the Alleghenies, and in 1763 and 1773 an elk. The elk in particular was touted as "well worth the attention of all lovers of natural history, and every other curious person." These curiosities from the interior of the continent joined malformed calves and dancing monkeys in the shows in New York without accompanying comment, making it hard to know exactly what their appeal was.[25]

While Britain had fought France in three previous wars, all of which had involved New Yorkers in at least some skirmishing with the French on their northern border, only with the Seven Years War that began in 1754 did New Yorkers become generally enthusiastic about defeating the French and their Indian allies. With the Seven Years War, eyes turned to the interior of the continent. Almanacs in New York, for example, suddenly printed maps of Canada, accounts of the war, and charts of distances to places like Pittsburgh. Perhaps these wild buffalo, huge snakes, and majestic elk stood for the wonders promised by expansion over the Appalachians, once the French were gone. But this in itself had nothing to do with refinement, nor did they represent God's serene natural world. Indeed the snake, which the report claimed had a small child and a dog in its belly when it was killed, promised a dark and unruly nature in the interior of North America, one difficult to contemplate with refined equanimity.

Even if shows of various curiosities in New York could not or did not reliably cultivate refined sensibilities in their audiences, reports of odd events did gradually take on the decorum of natural history in order to signify their truthfulness to their readers. The characteristics of such decorum, which grew out of long-developing skepticism over the content of traveler's tales and out of a desire to find a ground where men could converse without rancor, included careful and specific details, unemotional language, affirmation of the upstanding character of the witness or witnesses, and the absence of interpretation, thus skirting both a rush to judgment and controversial conclusions. Consider a 1745 story about a family who experienced stones spontaneously hurtling through the air inside their house, even after they had shut every door and window tightly. The witness, one of the many neighbors called by the alarmed family, gave the date (Nov. 29), place (Richard Davenport's house in Woodstock), a description of the stones' activity ("some flew with moderate Force between our Legs as we sat by the Fire; some fell gently into our Laps"), and a precise count of the stones (112). Nonetheless, the author forthrightly described this incident as "a surprizing Scene of the Preternaturals," interpreting it plainly as a ghost story. By the 1760s such stories were less likely to be interpreted. Here is one incident reported to have happened in Quebec in 1771:

> Tuesday, last, about 2 o'clock in the afternoon, the wife of one Joseph Beaume, at Lake St. Charles, being setting or working near the chimney, was unhappily killed by thunder.—Upon a strict examination on the body of this unfortunate woman, her head dress was found to be torn, and her head, tho' not deformed, as soft as a roasted apple; all the bones being calcinated, and as soft as cotton.—No other outward marks have been discovered.

Here are the plain facts: date, time, person, and material description. No hypotheses are ventured that might link thunder and a head "soft as a roasted apple." The incident is simply presented for the reader's information, about which he or she might draw his or her own conclusion.[26]

Even though newspapers reported on women spontaneously combusting, human figures seen walking calmly across water, and a knife swallowed by a cow and found years later harmlessly lodged "in her Brisket," readers were aware that some of these reports were disputed. One Richard Whitbourne of Corlaer's Hook wrote to a newspaper in 1762 in support of the existence of mermen. Although "some of the greatest naturalists" doubted their existence, Whitbourne said, he knew of reliable reports by participants in the siege of Louisbourg during King George's War, had read of sightings in well-regarded books, whose titles he gave, and knew that one was in a cabinet of curiosities in Denmark. If the oddities and "wonders" were no more absent from the natural world for most New Yorkers in 1770 than they had been in 1670, nonetheless, New Yorkers were conscious of a rising tide of skepticism and thus the need for more careful reporting and less confident interpretation.[27]

*I*f botany was of modest interest in colonial New York and electrical demonstrators had some difficulty distinguishing their demonstrations from the kinds of wonder shows that pleased the vulgar, the most prominent group to claim distinction based on their refined sensibilities toward the natural world were men with formal MD degrees. Like electrical experimenters, they claimed to be able to demonstrate the heightened mastery over natural phenomena and the selfless devotion to common good that were supposed to characterize gentlemen. As mentioned in chapter 2, in Europe from the Renaissance on, "physicians" were erudite men with university educations who diagnosed ailments and prescribed treatments, treatments that were then carried out by apprenticeship-trained surgeons and apothecaries. Thus physicians did mental work, while surgeons and apothecaries carried out physical healing work, a division that in theory affirmed physicians' place as akin to gentlemanly professions like law and the ministry, and placed the other two among skilled craftsmen. Even in Europe, however, this tripartite healing structure was common only among the wealthy at courts or in cities, and since few MDs migrated to Europe's colonies, it was largely irrelevant to the Americas. When a handful of MDs in New York in the 1760s and 1770s began to claim a heightened status based on their university education and their gentlemanly decorum, their claims were in principle familiar but by no means on that account easy to realize.[28]

Throughout the colonial period, New Yorkers continued to use the term "doctor" to mean any man who primarily earned his living through a general healing practice. A letter to the *New-York Evening Post* corrected a newly-arrived MD's assertion that a local apothecary wasn't a "doctor" but merely pretended to be one. The apothecary had "never as I have heard of stiled himself *Medicinae* Doctor," wrote the correspondent, "but as Doctor has been in most Parts of the World by Custom of Speech applied to all that belong to the Art of healing," the apothecary was properly addressed that way too. Men continued to learn healing through apprenticeships, by copying out medical books, and by availing themselves of the increasing opportunities offered by the expanding British military. Thomas Young, for example, born in the Hudson Valley in 1731, apprenticed himself locally and in Albany, and then had a peripatetic practice first in New York, then in Boston, and finally in Philadelphia. His cousin, Charles Clinton, enrolled as a surgeon's mate in the army in 1762. But even this much education wasn't necessary, for men of whatever training could set themselves up as doctors if other people would recognize them as such by paying them for their services. The traveling physician Alexander Hamilton noted in disgust in 1744 that a shoemaker on Long Island had a thriving doctor's practice because he had once happened "to cure an old woman of a pestilent mortal disease." A German physician who was stationed with British forces in New York City during the Revolution and then traveled through the new United States affirmed that "every man who drives the curing trade is known without distinction as Doctor, as elsewhere every person who makes verses is a poet." Such widespread and familiar social relations of healing were a distinct impediment to men who tried to claim an enhanced legitimacy based on erudition.[29]

Moreover, as with agricultural improvement, the expanding circulation of print, of commerce, and of people in the eighteenth-century British Atlantic reinforced the healing practices of ordinary households. With more newspapers and more almanacs, New Yorkers were offered more recipes for everything from ointments for chest coughs, to medicinal drinks for fevers, to gargles for sore throats. Thomas More's 1768 almanac, for example, offered recipes for the relief of disorders of the breast, scurvy, whooping cough, gout, dropsy, ague, rheumatism, and the bloody flux, along with his cider recipe, advice on hemp and flax cultivation, and directions for transplanting trees. Bookstores increasingly advertised domestic medicine books like Culpepper's *English Physician* and *The Lady's Dispensary, or, Every Woman Her Own Physician*. At least one version of *The Young Man's Companion* began to be printed with a supplement, *The Poor Planter's Physician*, that contained pages of recipes for household healing. In 1750 John Bartram wrote an appendix to Short's *Medicina Britannica* that specified what North American plants could be substituted for some of the plants in Short's compendium, and it was offered for sale in New York. Druggists in New York City offered "medicine chests" for sale, boxes with an array of commonly used medicines with directions enclosed suitable for rural farmers, such as they had long offered ship captains. As commerce accelerated generally in the colony, the number and variety of patent medicines advertised for sale in New York newspapers rose, strikingly so after 1750. This efflorescence of recipe knowledge and of access to medicinal substances notably expanded the repertoire of healing among ordinary New Yorkers.[30]

Just as printers, booksellers, and apothecaries offered opportunities to expand household healing, other people found new opportunities in healing work, just as smallpox inoculators had. Some offered nostrums for sale, medicines they made themselves that promised to cure one ailment or another. Thus did Bartle Murfoy in Westchester County offer a salve that cured rheumatism, sciatica, gout, or other joint pains "in less than Four Hours Time with the Assistance of God." John Levine in New York City offered his "eye water" for sale, and William Clark in New Jersey advertised an oil that eased urinary stones. Women, who in New York were as strong a presence in commerce as anywhere in the British Atlantic, also offered nostrums for sale, as Elizabeth Richards did with her salve for sore nipples or Margaret Elkin with her eye water. While there is only a passing mention of one woman in New York, a Mrs. Katherine Howard, who seemed to function fully as a doctor (her obituary described as a "Professores of Physick"), women regularly healed for pay as midwives and nurses. But note that when Mary Richardson advertised her skills, she began with female ailments like sore nipples, proceeded through similar ailments like running sores, and went on to unrelated ailments like throat disorders and pain in the limbs. In her ad one can see that the boundary between women's and men's healing work was considerably blurred. And whether reporting new discoveries in healing or their own services, people consistently secured their claims with the testimony of satisfied customers or the promise of "No Cure, No Pay." Refinement was irrelevant here. As with medicinal recipe advice, what people could experience for themselves continued to be the touchstone of reliable knowledge.[31]

Nor was there a clear boundary between erudite and modest origins for healing practices. In 1739 the dog of a Westchester farmer was bitten by a rattlesnake. Having read in the *New-York Gazette* that the Royal Society in London had performed an experiment using salad oil as an antidote to poison, the farmer used it to cure the dog. *The Husband-man's Guide*, a considerably less erudite source, also recommended salad oil but said hog lard might serve as a substitute. In 1743 a Hudson Valley farmer and part-time doctor, Evan Jones, discovered one of his cattle with his head "swelled to an Incredible bigness." Jones suspected that the animal had been bitten by a rattlesnake and, having no salad oil at hand, he melted down some hog lard, poured some down the steer's throat and smeared the rest on its head. The hog lard, Jones reported, was a success. He then made his own experiment. He had an employee catch a rattlesnake, and prodded it to bite one of his chickens. When the chicken was reeling from the effects of the poison, Jones treated it with hog lard, again apparently with success. "From this we may learn," Colden, who was Jones's neighbor, wrote, "not to despise Remedies because common and which for that reason appear to be mean."[32]

In this context, doctors continued to operate within what Nicholas Jewson has called "a force field of patronage," their opportunities constrained by people's willingness to give them their business. As adjuncts to household healing, doctors were called in when people felt they needed additional advice or specific skills such as setting bones or letting blood. In the first century of colonization, doctors of any sort were reasonably sparse in New York, but as the population rose, so did the numbers of men trying to make a living as doctors. By the 1750s competition among them was increasingly fierce, and was conducted much like commercial and political conflict in the colony, that is, very personally. Letters to newspapers detailed the malpractices of rivals with identifying details only thinly veiled. In October 1751 a letter appeared in the *New-York Weekly Post-Boy* reprinting a prescription for a plaster written "for Mrs. S——ple, signed C.L.S," a prescription, the writer claimed, that included ingredients that didn't exist and in proportions that wouldn't yield the appropriate consistency. "In this extraordinary Performance, there is total ignorance of Surgery, Pharmacy, and Physick," proclaimed the writer. The following week Charles Leslie Schaw, an MD newly arrived in New York, protested that he had been impugned, and although responding to such public aspersions was "indelicate" in a gentleman, nonetheless he felt called upon to defend his honor and his skill as a physician. But his rivals, including William Brownejohn, a doctor trained by apprenticing to an apothecary and who had practiced in New York City apparently to general satisfaction for fifteen years, prevailed in the court of public opinion. Despite testimony from the husband of "Mrs. S—ple" that his wife had recovered perfectly well under Schaw's care, Schaw found it necessary to relocate to New Jersey.[33]

A common weapon in this competition among doctors was to asperse a rival as a "quack," an accusation that could and did lead to dueling in the streets. "Quack" was a general smear against a claim to competence and not, as we would have it, a term indicating the absence of a right to practice. As such it was applied to doctors with and without formal medical education. Thus Alexander Hamilton called a drunken doctor he encountered in New York City who had no university education or apparently any familiarity with medical books, a quack, and James Jay, who had a perfectly good MD degree from the University of Edinburgh, was described in a private letter as having "the true principles of a Quack." Moreover, people whom we would now associate with quackery evidently drew no such opprobrium if their healing skills seemed satisfactory. In 1771, for example, the *New-York Journal* described an itinerant, Dr. Anthony Yeldell who had set up a stage in

Brucklyn and "who by his Harangues, the odd Tricks of his Merry-Andrew, and the surprising Feats of Activity of his little Boy, highly diverts the People." Here is our image of the shady mountebank, but the newspaper, other than distancing itself from Dr. Yeldell by referring to him as a "stranger," merely described him as the backdrop to the real story, the wreck of the overloaded ferry returning to Manhattan. Three years later Dr. Yeldell had moved on to Philadelphia, but advertised in New York to his old customers. People's perception of doctors' skills and sense, rather than of doctors' formal credentials, was the basis for a viable healing practice.[34]

This was the background, then, against which some men worked to persuade New Yorkers that an MD degree distinguished physicians from the common run of healers and entitled them to a higher level of respect. Primarily, their argument was that improvements in learning, dating from Harvey's delineation of the heart and the circulatory system published in 1628, had so deepened the understanding of the body and disease that only the erudite could be relied upon for medical guidance. Indeed, European physicians from the late seventeenth century on had worked to apply the standards by which knowledge was secured among natural historians and natural philosophers in order to discover the natural laws by which the body worked. Some tried to make taxonomies by which fevers could be categorized in order that each species of fever could be treated distinctively and appropriately. Others tried to apply developments in mechanics to medicine, redescribing the body in terms of its irritability and elasticity or the firmness or laxness of its solid parts. More attention was given to anatomically distinct parts, like the liver or the nerves, in contrast to the more generally undifferentiated body of earlier centuries. While most diseases continued to be conceptualized as a derangement of the body as a whole, rather than as ailments specific to particular organs or parts, nonetheless these efforts represent attempts to apply a more critical and analytic approach to illness, in tandem with a rejection of ancient models. A letter to the *New-York Weekly Post-Boy* in 1767 praised medicine for "beginning to take an *experimental* Turn." Making medicine the province of learned physicians, promised Peter Middleton, an MD, in 1760, would offer "the Pursuit of farther Discoveries in that useful Profession; which in all Times, and among all polite Nations, has ever been esteemed Honorable, and worthy of Men of the first Rank and Learning."[35]

That such erudition should go hand in hand with a refined and elegant comportment, the comportment supposedly natural to "men of the first rank," was understood. The experiences of the Bards, father and son, demonstrates how one New York family hoped to raise their fortunes through investing in a university education through which the son would both become an MD and attain a gentleman's polish. In 1761 John Bard sent Samuel to the medical school at the University of Edinburgh, among the most prestigious medical schools in Europe. John Bard was a doctor himself, but apprentice-trained, and while comfortable and respected, he wanted more for his son. Father and son took pride in Samuel's medical studies and his attending the lectures of eminent professors. Nonetheless, John reminded Samuel to pay attention to "appearing like a Gentleman," urged him to converse with young ladies when given the opportunity to improve the ease of his manners, and approved of his desire to take a class in "Rhetorick and the Belles Lettres," which, Samuel wrote his father, "altho' no physical class, I am far from thinking unnecessary to the character of a gentleman." Samuel was the young man mentioned in chapter 4 who investigated madder cultivation for his father. But Samuel also gave his father advice on laying out fashionable gardens and recommended a program of reading to his brother that will "refine your taste and add a Delicacy to your Sentiments and Manners." John Bard was well pleased with the progress his son was making in both his medical studies and his refinement. "[A] Gentleman will make his Practice Valluable in Proportion to his Merrit and address," John wrote Samuel. "[A]n Easie, obliging, and attentive manner with a proper degree of Skill will Insure you the best practice and the highest fees."[36]

Physicians in New York tried to do more than simply rely on their reputation for learnedness and their polish as gentlemen to raise their standing. One strategy was to institute medical licensure, with a group of formally educated doctors examining applicants. There were calls for licensing physicians as early as 1749 (midwives in New York City had been licensed for more than ten years by then) and a licensing board was established in 1760. Medical education too was improved. In 1763 Dr. Samuel Clossy, with a medical degree from Trinity College, Dublin, arrived in New York City and, finding it difficult to establish a sufficiently lucrative practice, began to offer anatomy courses. New Yorkers like the Bards heard enviously of the new medical school in Philadelphia; in 1767 they established one in New York City under the auspices of King's College. Samuel Bard was the first professor of physic, Clossy the professor of anatomy, and four others assumed the positions of professors of physiology, surgery, material medica, and midwifery. In 1770 the medical faculty persuaded the Assembly to fund a hospital so students could study unusual illnesses (for, one admitted, "every sagacious and attentive old Woman" could make judgments about the usual ones").[37]

While this handful of physicians were able to institutionalize learned medicine, their actual effect on healing practice in New York was modest. Few men applied for a license to practice medicine (and no one was disciplined for practicing without it), only a handful of students graduated from the medical school before the Revolution intervened, and while the hospital was built, it burned down almost immediately. Worse, from the point of view of doctors like the Bards, the newly arrived or newly minted MDs were wont to thrust their erudition before the public in a manner that was less than gentlemanly. John Bard complained that some of them "cannot keep their profession out of their heads a moment, and one of them dashing ones brains out with Technical terms on all Occasions . . . Where the Physician and the Gentleman unite," he added, "you never perceive anythin of this—it argues both Ignorance and Vanity—not perhaps Ignorance of his Profession, but of Everything Else." Vaunting their learning was less like the effortlessness expected of gentlemen and more like the fractiousness of a pedant. If the weapon these physicians had in their competition with the host of other healers was their erudition, wielding it proved double edged.[38]

Far from inspiring admiration among New Yorkers, the erudition spouted by these new MDs in fact roused their suspicion. Certainly, New Yorkers resisted the theoretical distinction of physicians as merely diagnosing and then directing surgeons and apothecaries to carry out their prescriptions. As the merchant John Watts commented dryly, "People in America are very fond of good Pennysworths, they choose to have the Physick and the Advice go together, and had rather be work'd to death than give their Money for simple Advice." But beyond that, there was a widespread conviction that erudite language was merely a ruse to gull the unlearned. "A Dissertation upon Corns in a Letter to a Friend," a satire printed in the *New-York Weekly Journal*, mocked the propensity of learned medicine to describe common ailments in arcane terms. And the use of physicians by New Yorkers coexisted with persistent suspicions that they did not, in fact, do much good. As a ditty in a newspaper put it,

> One prompt Phisician like a sculler plies,
> And all his Art, and skill applies;
> But two Phisicians, like a pair of Oars,
> Convey you soonest to the Stygian Shores.[39]

Moreover, the claims of doctors—MDs or not—to benevolence towards the sick were undercut by their undeniable financial stake in accruing business. A letter to the *New-York Weekly Post-Boy* criticized local doctors for not suggesting tar water, a common medicine made by letting pine resin sit in water. "If then Physicians should prescribe Tar Water, or such Things that could not be disguised from the Knowledge of the Patients," doctors would earn correspondingly small fees, the writer pointed out. Because the household continued to be the site of most healing, because people believed that in general they understood their own bodies and the sources of its derangements, and because people had an increasing array of homemade remedies and purchased compounds they themselves could use in healing, men with claims to learned understandings of health and disease had difficulty persuading their fellow New Yorkers of their usefulness.[40]

And yet by the latter half of the eighteenth century, both the practice of healing and the understandings of the body had shifted somewhat. The practice of medicine no longer looked back towards ancient models but forward, striving to participate in the more general eighteenth-century process of improvement. Some procedures were genuinely improved, like lithotomy (operating for urinary stones) and cataract removal. In New York, the most prominent lithotomist was John Jones, the son of the doctor-farmer who had used hog lard, who had earned an MD from the University of Rheims and became the professor of surgery when the medical school opened. Jones repeatedly and successfully performed lithotomies, operations reported admiringly in the New York press. Nonetheless, his success wasn't particularly attributed to his erudition but to his skill. Similarly, the "oculists" who brought cataract removal to New Yorkers, and who were not MDs, were admired for what they could do, not for what they could say. Thus while erudition in medicine drew little admiration among New Yorkers, nonetheless there was a widespread sense that the actual practice of medicine—by skilled people or by new recipe cures—affirmed a belief in the more general progress in knowledge.[41]

If erudite medicine was little regarded, nonetheless ideas about the body that it advanced crept into New Yorkers' vocabulary. Even though no one questioned that most ailments were derangements of the body as a whole, the older conception of the body as a sac of fluids, with the balance of four humors as necessary to health, was modified. For one thing, black and yellow bile generally disappeared altogether from descriptions of illness, replaced by references to generalized "humors," or sometimes just "juices." A powder advertised in 1759, for example, promised that it would "purge the Humours" and another for Turlington's Balsam of Life lauded it as "the most powerful corrector of Juices ever before prepared." Phlegm continued to be noted on occasion but increasingly took on its modern meaning of the substance coughed up from the lungs, rather than a fluid that circulated throughout the body. Of the four humors, only blood remained prominent and, indeed, remained of prime concern. While purging the body of excessive or corrupted humors in general by bleeding, inducing vomiting or stool, or sweats and blisters continued to be the mainstay of treatment for illness, maintaining the right consistency of the blood and purging it of impurities figured large in healing practices.[42]

Although concern with bodily fluids continued to be important, there was a sharpened awareness of new concepts in anatomy. A "Female Cordial" was advertised in 1769 as "peculiarly adapted to the delicate nervous Texture of the Female Sex," and another ad in 1772 touted American Balsam as excellent for people with "loose fibers." In 1774 "Roger Meanwell" printed a new version of the Man of Signs, still nearly ubiquitous in New York almanacs, this one with the abdomen opened so that the internal organs were on view, a reflection of the enhanced understanding of the significance of distinct body parts. New Yorkers continued to conceptualize the skin as porous in both directions, allowing excessive bodily fluids out and noxious air in. But their sense that each body was constituted differently and was so affected by shifts in the external environment such that "the very Medicine that is of Service one Day . . . may be destructive the next" was implicitly contradicted by their embrace of patent medicines. Patent medicines, after all, promised to heal some given ailment no matter in whose body it appeared or under what circumstances. To some degree then, the claims of erudite medicine to a more accurate understanding of the body as a mechanism and as one in which the fine points of anatomy mattered did alter older understandings of the body. But these shifts in understanding were subtle, and since they were compatible with familiar social relations of healing and with the incorporation of new healing techniques, they occasioned little sense of dislocation among New Yorkers.[43]

To the degree to that anything "refined" about healing was adopted by New Yorkers, in contrast to the merely "improved," it was in a heightened emphasis on "temperance"—working to keep the body in balance by the moderate ingestion of appropriate foods and the moderate exposure to the elements and to exercise. Refinement as a social ideal incorporated restraint and moderation, such as the "Regulation of our Passions as is most conducive to the Wellbeing of the Animal Oeconomy." New Yorkers were warned against immoderate drinking of alcohol, mourned the death of loved ones "of a fever occasion'd by a surfiet taken by over Exercise," and read repeatedly of people who dropped dead after they drank cold water in hot weather. But temperance had long been a mainstay of health in the early modern world, and the degree that exhortations to act temperately were more visible in 1770 than they had been in 1670 may well be merely an artifact of the considerable expansion of print.[44]

In any case, despite the efforts of men like the Bards, doctors got little credit for bringing refined understandings of the natural world to New York. In fact, one of their key practices, that of dissection, considerably undermined their claims to being gentlemen. By the eighteenth century, dissections affirmed the body as a mechanism and allowed physicians to probe for knowledge beneath the skin. As a poem, "On the Dissection of the Body," expressed it,

> Though God has call'd the life he lent,
> Each vital function, dormant laid,
> Here we trace nature's deep intent
> And see how once the springs were play'd.

In their own minds, physicians' contemplation of bodies was akin to the South Carolinian's contemplation of lightning: men informed by understanding could gaze serenely at what struck fear and horror in ordinary people. But in New York, as elsewhere in the Atlantic world, physicians who tried to obtain bodies for dissection met considerable resistance. Samuel Clossy wanted bodies for his anatomy lectures and complained that he and his assistants had become so well known "that we could not venture to meddle with a white subject and a black or Mullato I could not procure." In New York as elsewhere, autopsies had long been performed when there was a reason to suspect foul play as a cause of death. But even these were resisted. When in 1767 Sarah Parker, the wife of a sailor out at sea, was found dead in her bed, the sheriff's deputies had to overcome five of her women friends who barred the door to their entrance. They emphatically did not share an understanding of the dead body as a source of insight into and mastery over the natural world.[45]

Partly this was because dismembering bodies was long associated with the treatment of criminals, or, as Clossy's complaint implies, with an inability of the deceased to command respect. But it was also because New Yorkers understood that the line between life and death was sometimes indistinct. An occasional newspaper story told of people thought to be dead who awakened. Others explained how to revive people who had drowned or suffocated. In fact, Dr. Clossy was among those who gave such advice to his fellow New Yorkers. And one of the ways in which murderers were discovered was by having suspects approach the murdered body, which, it was widely believed, would respond to the presence of its murderer. In such a case in 1767 in Bergen County, New Jersey, the jury of inquest charged a man with murder when the corpse bled from the nose on his approach. In this understanding of death, uneasy bodies retained some memory of lived experience and so could still "speak." Physicians, then, who tried to claim dissection as a practice that affirmed their calm mastery of the material world could be understood not as having a refined sensibility but rather not having much sensibility at all.[46]

*I*n the end, refinement was only a modest entry point for new realms of knowledge about the natural world for New Yorkers. If wealthy merchants, landowners, and lawyers in New York embraced elegant consumption, few extended their interests to refined sciences like natural history or electricity. Some prominent New Yorkers did embrace the restraint and decorum associated with refinement. The lawyer William Livingston helped establish New York's first lending library, was the major contributor to a newspaper modeled on *The Tatler,* and penned a poem lauding the beauties of a modest and rural life. But there were New Yorkers of lesser wealth and standing, like Mary Cooper, the Long Island farmwife mentioned in chapter 4, who similarly struggled to lead lives of virtue and restraint. By contrast, plenty of New Yorkers high and low enjoyed cockfights, horse races, and Pinkster festivals. As a cultural style, then, refinement did not separate the gentle from the vulgar in New York, as people continued to rub elbows freely, if contentiously, up and down the social scale.[47]

Nor did refinement come to signify character reliably. In the 1740s a young man introduced himself in the West Indies as a Livingston from New York. He was embraced by a number of wealthy families on the basis of the Livingstons' reputation and his own refined deportment. When, however, he tried to make off with one family's valuables, he was revealed as plain Tom Bell, a sea captain's son from Boston, and his escapades were reported in New York newspapers. The appearance of refinement, then, was no assurance of authenticity. And in any case, politics and commerce in New York remained so competitive that, as with the doctors, personal alliances and rivalries continued to structure access to credit and office, not the presentation of some gentlemanly ideal. With refinement of only modest significance among a wide swath of New Yorkers, it is unsurprising that new modes of refined knowledge of the natural world found little purchase here.[48]

Nonetheless, as we have seen, some New Yorkers found deep satisfaction in these new modes of understanding the world and were able to refashion themselves through them. Certainly Cadwallader Colden found friendship and new sources of esteem through his pursuit of botany. In the world of his correspondents he was a worthy gentleman, even if he wasn't much respected in New York itself. Jane Colden too evidently found pleasure and most likely an enhanced importance in her mastery of botany. Samuel Bard became the respected physician his father hoped he would become. A man dedicated to furthering the development of medical knowledge, and evidently of a kind and obliging temper, Bard wrote his father proudly in 1773 that he had been asked to join the Church and King Club, a gathering of gentlemen. And if New Yorkers who took these pursuits seriously were relatively rare, refined knowledge of the natural world had become fashionable, if only superficially. On the Mohawk frontier, the trader William Johnson, now made a baronet through service to the Crown and grown rich through land speculation, laid out a fashionable garden, ordered a telescope, protected his new mansion with lightening rods, and deplored the fact that a physician with a "liberal education" couldn't get enough business in his area to make it worth settling there. By the 1770s William Young of Oyster Bay advertised his nursery as a "Linnaean Botanic Garden," booksellers like Samuel Loudon offered titles like Captain Cook's *Journal of a Voyage Round the World,* and almanacs routinely printed pretty verses like the one that heads this chapter. If only a few used the new sciences in their own refashioning, fashion at least made them familiar to many more.[49]

Some historians have argued that the rise of science contributed to a profound shift in human consciousness, one in which people increasingly became alienated from the natural world. A natural history that severed each plant and animal from local interrelationships in order to place them in a universal catalogue rendered those interrelationships invisible and hence increasingly negligible. Explanations of hitherto mysterious forces like lightening bleached the natural world of spiritual force, rendering it inert. Reconceiving the human body as a complex mechanism withdrew it from the comprehension of ordinary people and made it the province of specialists. However much this alienation may characterize modernity, these changes in the knowledge of the natural world penetrated among ordinary people very slowly and made little difference in the short run. If New Yorkers, in print at least, were more reticent about interpreting odd phenomena, nonetheless many plants around them remained wild and unnamed, twenty-one-foot snakes swallowed small children, dead bodies accused their murderers, and their own bodies still oozed and stank. Natural historians, electrical demonstrators, and physicians might claim to be en route to penetrating the laws of an orderly universe, but the nature of ordinary people continued to be capricious and unruly.

CHAPTER 6

Reason

What tho' in solemn Silence all,
Move round the dark Terestial Ball;
What tho' no Voice, nor real Sound,
Amid their radient Orbs be found.
In Reason's Ear they all rejoice,
And utter forth a Glorious Voice,
Forever Singing as they Shine,
The Hand that made us is Divine!

Thomas More, *Poor Thomas Improved.*
Being More's Country Almanack for . . . 1768

*I*n 1712 Robert Hunter, the governor and Fellow of the Royal Society whom we met watching Morgan maneuver his boat and persuading Colden to move to New York, ordered a census of his colony's inhabitants. He was, however, thwarted by "the people being deterr'd by a simple superstition and observation, that the sickness follow'd upon the last numbering of the people." New Yorkers had apparently drawn a conclusion from their experiences that they judged reasonable, but Hunter did not. Three years later Hunter gave up. "The superstition of this people is so insurmountable," he wrote the Lords of Trade, that he was reduced to approximating the population using lists like militia rolls and slave enumerations. When Hunter drew his particular line between the reasonable and the superstitious he neglected to point out that New Yorkers were not merely heeding their own experience. In the Old Testament David commanded a census in order to display his worldly power, and as punishment for David's pride God had visited pestilence upon the Israelites. Thus the material experiences of New Yorkers with their last census were linked to their reading of this biblical text (which may also have had political meaning, for David's error was his governmental hubris). While Hunter may have deemed New Yorkers' resistance superstitious, they themselves were acting upon reasonable conclusions drawn simultaneously from biblical teachings and their own material experiences.[1]

Reason and superstition are rhetorical opposites. People who think themselves reasonable claim that they see the world as it really is, use logic to make connections between cause and effect, make dispassionate observations, and come to intelligent judgments. They see the superstitious as irrational: claiming experiences that are materially impossible, subjecting themselves to baseless fears, and drawing erroneous connections between unrelated events. This contrast between reason and superstition took on a particular salience in the eighteenth century, as some people worked to shift some beliefs that were previously common sense out of the reasonable and bundle them with the superstitious. Like improvement and refinement, this heightened emphasis on reason both offered new ways of understanding the natural world and increasingly served to mark the social boundary between the genteel and the common.

In the seventeenth century most people in New Netherland and New York took it for granted that the Devil was active in the material world, neighbors could become witches, shifting alignments of planets in the sky affected weather on earth, and comets were portents of God's judgments. In the eighteenth century "reasonable" people came to believe that the Devil was merely metaphorically active, witchcraft was materially impossible, planetary movements were irrelevant to events on earth, and comets were natural objects whose orbits were mathematically predictable. At the same time, efforts to apprehend the natural world more reasonably did not bleach natural phenomena of their spiritual significance. While in the nineteenth century and even more strongly in the twentieth, the reasonable apprehension of the natural world would come to occupy a cultural space distinct from religion, in the eighteenth century British Atlantic there was no such distinction. Few in the eighteenth century seriously disputed that the natural world was a text of God. And if it was a text of God, it was also a reflection of his nature and of his intentions towards mankind.

Some new assertions about the relationship between God and the natural world, such as botanists' avowal that studying the beauty and order of plant life lifted one's eyes to God in worshipful reverence, were anodyne enough to rouse little controversy. But other assertions about the natural world were much more at odds with many people's theological beliefs and consequently raised fervent opposition. Assertions that emphasized how the natural world followed natural laws, for example, contradicted many people's understanding that odd or unexpected events were religious messages. In particular, new assertions about astronomy and cosmology, because they raised fundamental theological issues and because they raised profound questions about New Yorkers' understandings of their everyday material experiences, were often vehemently contested.[2] .

Even if Hunter couldn't persuade New Yorkers to cooperate with his census in 1712, over the ensuing decades the boundary between reason and superstition clearly shifted. Like news of improvements and representations of refinement, essays and stories in newspapers everywhere in the British Atlantic disseminated new distinctions between the rational and the foolish. Some New Yorkers exerted themselves to spread reason's gospel among their fellows, as Hunter had tried to do. If in the seventeenth century accounts of witchcraft, ghosts, magic, omens, and messages in dreams seemed plausible, these were now all increasingly ascribed to the ignorance of the vulgar. Yet in the end, while these efforts raised the level of skepticism in the colony, the boundary between reason and superstition was more a shifting borderland than a clearly marked line, for some New Yorkers remained unconvinced by the new definitions of reason.

*L*etters and essays in newspapers increasingly scoffed at beliefs in manifestations of magic and the Devil. In 1729 Samuel Johnson, an Anglican minister in a Connecticut town about thirty miles from New York, published a series of essays in the *New-York Gazette*. His intent was, he wrote, "to free and manumit Mankind from the many Impositions, Frauds, and Delusions which interrupt their Happiness, so I shall attempt to remove the Popular Impressions and Fears of Spirits, Apparitions, and Witches." He contrasted truly understanding the changeability of the natural world with the ignorance that underlay misapprehension. Thus, "Great Guns were esteemed by the Americans to be angry Deities; Ships, floating Monsters; the Sun to be the God of the World; Watches to be living Annimals; Paper and Ink to be Spirits, which conveyed Men's Thoughts from one to another; and a Dancing Mare was lately burnt for a Witch in Portugal." Johnson's abrupt shift from evidence of Indians' ignorance to the actions of the Portuguese made common beliefs about the actions of the Devil in the material world the equivalent of heathenism. When God sends messages, Johnson declared, they are clear and public, not the secret province of "Miracle-Mongers." Stories of goblins or old women on broomsticks yield no lessons for mankind, and therefore they cannot be part of God's design for the natural world. Nor could naturally occurring events demonstrate the power of the Devil. The Devil plays no "Monkey Tricks in Church-yards," claimed Johnson, for we all know he was cast out of heaven and chained in the regions of the damned for eternity. Similarly, an anonymous essayist declared in 1751, the belief in

ghosts is due to a "timorous Imagination" or a "low and vulgar education."
In these stories and essays, witches, ghosts, and the Devil's actions on earth
dwelt only in the imaginations of the ignorant.[3]

When Johnson targeted "frauds and Superstitions" in his essays,
his larger purpose was to persuade his readers that when they believed in
"Prophets, Fortune-Tellers, Conjurors, Witches, Apparitions, and such like
Superstitious Fooleries," they not only misapprehended the nature of the
natural world but also the nature of God. As with his example of the Portuguese mare, the charge of superstition was widely used by Protestants to
distance themselves from Roman Catholicism. A newspaper report in 1731,
for example, described the appearance of a new star over a town in France,
which caused much consternation and predictions of calamities until it was
discovered to be merely a lantern on the tail of a kite flown by a "roguish
English boy." Another report in 1760 of Catholics parading a Saint's image
in response to an earthquake noted sarcastically that "it had not the desired
Effect, many severe Shocks happening afterwards." Catholics, widely de-
spised by Protestants in Britain, the Netherlands, and the colonies, epito-
mized people who were subject to "frauds and superstitions."[4]

Despite this widespread consensus that Catholics misunderstood the
natural world because they were ignorant of the nature of true religion, in
New York as elsewhere these sorts of slurs were also used as weapons in
intra-Protestant conflicts. Johnson himself declared that of all the "Superstitious Fooleries" common to mankind, the very worst was "enthusiasm."
Within eighteenth-century British Protestantism, the term "enthusiasm" was
used by adherents of churches with stronger commitments to rite and clerical authority to disparage those who emphasized personal conversion experiences. Johnson had himself been brought up within New England Calvinism
but converted to the Church of England while a student at Yale. His essays
were part of his efforts to raise the visibility and the legitimacy of Anglicanism in a region where it commanded only a modest number of adherents.
Yet twenty-five years later, when Johnson was offered the presidency of
the newly-established King's College in New York, it was his Anglicanism
that an opponent, Presbyterian William Livingston, claimed would subvert
the development of reason among students. "Where, instead of Reason and
Argument, of which the Minds of the Youth are not capable," Livingston
wrote, "they are early imbued with the Doctrines of a Party, inforced by the
Authority of a Professor's Chair, and the combining Aims of the President."
By contrasting the "Miseries of Priest-Craft" with the "Liberty wherewith
Christ has set us free" in arguing against the establishment of a "Sectarian"

college, Livingston tarred the Anglican Johnson with the brush of superstitious Catholicism, a rebuttal to Johnson's assertion that it was "enthusiastic" Protestants who were superstitious.[5]

It is difficult to map specific beliefs about manifestations of spiritual force in the natural world onto formal denominational allegiances. For one thing, essays and letters published in the newspapers were often anonymous, making it difficult to know the religious allegiances of the writers. For another, beyond the general tenet that the natural world was a text of God, nowhere in Britain or its colonies did different denominations take specific formal positions on particular assertions about the natural world. Indeed, some denominations divided internally on such matters. In Britain, for example, the latitudinarian wing of the Church of England, which rejected imposing strict doctrinal and liturgical uniformity, prominently embraced a vision of an orderly universe obeying natural laws, thus demonstrating a benevolent God. But the Laudian or Tory faction was often opposed to reading the word of God in the book of nature altogether, believing that the knowledge of God came solely through scriptural sources. Some Presbyterians like the electrical demonstrator Ebenezer Kinnersley in Philadelphia were as committed to a lawful universe as any latitudinarian and opposed the "enthusiasm" he saw overtaking other Presbyterians during the Great Awakening. Still others, like the minister Henry Melchior Muhlenberg, who occasionally ministered to congregations of German Lutherans in the Hudson Valley, were generally indifferent to the natural world altogether, finding like some Laudians nothing that he needed to know there. Muhlenberg once attended the electrical demonstrations given by Kinnersley. He was unimpressed. "Great are the Lord's works of creation and preservation," he commented in his diary, "but even greater is the work of redemption!" And thirdly, assertions about what constituted "reasonable" understandings of the natural world were so diverse and variegated that very few things were placed firmly and consistently on only one side of the border between reason and superstition.[6]

Among the diverse strands of thought that circulated in New York and complicated any straightforward description of what was reasonable were Neoplatonic and Paracelsian ideas about the "living" nature of seemingly inanimate objects, the unseen connections between disparate phenomena, and the ability of matter to shift from one form to another. Despite the consistent assertion that "experience" was the best test of the truth about natural phenomena, a conviction found high and low among New Yorkers, the inexplicable was sometimes attributed to just such intangible connections. "Sympathy," some sort of innate correspondence that operated at a distance, was one such example. In 1752, for example, Thomas More offered a recipe for getting rid of caterpillars in apple trees by placing sod in the crotches of the trees, which caused the caterpillars to fall to the ground. "This is said to be done by Sympathy, and told for Truth by one that try'd it," wrote More. Roger More in his 1761 almanac gave a recipe for "Roman or Hungarian Vitriol," good for the cure of wounds because it would "attract the corruption of his wounds, and stay the bleeding of them," even if the patient were miles away. The printer James Parker advertised an "anodyne necklace" in 1747, whose beads prevented the death of children thought past recovery due to teething, fits, fevers, or convulsions. And in 1772 one New Yorker wrote another of his great satisfaction that the nosebleed of a friend had ceased "by means of a Necklass of Amber Beeds I put on his Neck." Pregnant women were widely known to impress particularly vivid experiences on their fetuses. The *New-York Weekly Post-Boy* reported, for example, a case in Hartford, Connecticut, in which a pregnant woman was seized with a longing to eat the flesh of her husband's arms. While the husband tried to oblige her, her teeth were not sharp enough to pierce his skin. Nonetheless, when the baby was born, it refused to take anything until it was offered raw fowl flesh, of which it ate heartily. Samuel Johnson, too, gave examples of phenomena that seemed to demonstrate such mysterious connections, such as how "The Lodestone draws Iron to it; Gold attracts Quicksilver; The Sensitive Plant shrinks from the Touch" and a "Rattle-snake by a sort of Magical Power in his Eyes will force a Squirrel into his Mouth." These, too, were evidence of an orderly nature, for these phenomena were consistently observed by reasonable people, even if the laws by which they worked were not altogether clear.[7]

Nonetheless, when Johnson offered his list, he offered these items as phenomena that were "impenetrable to our vain and fruitless Enquiries." This, too, was a strand of thought that complicated what constituted "reason." Johnson was doubtful that the human mind could penetrate very deeply into the ways in which the natural world worked and further, he doubted that material experiences were always a reliable guide to such understanding. After all, he said, "straight Things in the Water appear Crooked" and "All Things appear Yellow to Men in the Jaundice." His doubts about the reliability of sensation and the weakness of human understanding were sometimes echoed by other New Yorkers. In 1752 the almanac maker Roger Sherman observed

Let Sages search God's Works with Ard'ous Scan,
They are past finding out by mortal Man.

Like Daniel Leeds before him, Sherman believed he could offer only modest assurances of what the natural world was really like.[8]

At the same time, despite the commitment to an epistemology of concrete experience and this antiphony of skepticism about the reliability of sensory information, some New Yorkers felt free to speculate in areas where no material experience was available. The idea that celestial bodies like the moon might be inhabited was proposed periodically by a number of people in Europe, like John Wilkins in the 1630s, who would be one of the founders of the Royal Society. Titan Leeds printed a rhyme in his 1732 almanac in which a farmer pulled out his copy of Wilkins to prove that the moon was inhabited, whereupon his companions fall silent, happily imagining the trading opportunities an inhabited moon would offer America. Johnson, too, thought it "very likely that Earth's Fellow-Planets, who move about the Sun (as we do) are fill'd with Inhabitants." An essay, "On Reason and Science," printed in the *New-York Gazette* in 1732 believed that the stars nourished other inhabited planets, unknown to us, for God "does nothing in vain" and so must have made the stars for some purpose, a belief echoed by Jesse Parsons in his almanac of 1757. In his 1773 almanac Roger More rhymed

> Yet reason says, nor can we doubt at all,
> Millions of *Beings* dwell on either ball,
> With constitutions fitted for that spot,
> Where Providence, all wise, has fix'd their lot.

Reason, then, led many New Yorkers to believe in phenomena that were out of the reach of their senses altogether.[9]

As with wonders, anomalous natural phenomena could be interpreted providentially or not, another complicating factor in drawing a sharp line between reasonable and superstitious understandings of nature. A report of a 1750 earthquake in London drew this aside in the *New-York Weekly Post-Boy*: "Happy! would it rouze them from their FOLLIES and INIQUITIES." Similarly, news of the great Lisbon earthquake of 1756 led a reader to submit two poems to the *New-York Mercury*: "The Soliloquy of an Impenitent" and "The Soliloquy of a true Christian wak'd by the Earthquake." In one, the poet has the Christian cry

> Be still, my Soul, and Pause!—
> I hear the Shrieks of Horror and Despair:
> 'Tis so! It's Nature's GOD, bids Nature be no more.

The providential meaning was not always so explicitly interpreted; some earthquakes were merely reported and other reports were accompanied by un-interpreted observations of accompanying phenomena. Cadwallader Colden, for example, wrote a description of an earthquake in November 1755 in which he pointed out that the previous summer and fall had been the driest ever known and the sky was unusually clear for several days prior to the quake, while smoke drifted downward instead of up and his watch had run unusually slowly. In the Mohawk Valley, the Anglican minister John Ogilvie recorded the same earthquake in his diary, but added that people reported to him that they saw "several Balls of Fire in the Air passing very quickly" accompanying the earthquake. Colden, then, recorded only what seemed to be purely natural phenomena, while people in the Mohawk Valley observed phenomena generally associated with providential messages. And other celestial phenomena, like the aurora borealis around which there was no providential interpretive tradition, were simply reported with admiration.[10]

Of all the phenomena about which people debated, the ability of the Devil to seduce people into witchcraft and in turn witches' ability to perform magic was the one most decisively placed over the line as superstitious. New York newspapers periodically reprinted stories from the British press that were all variants of the same message. In these accounts, "country People . . . full of Ignorance and Superstition" accused some poor old women of being witches, and then "Gentlemen" fortunately intervened (although not always in time.) Other stories told of the deception of credulous people, as in one of a woman crying in the street that the Devil had hold of her. The "Devil" turned out to be an old black sow and the only harm to the woman, reported the story, was some "genteel perspiration." In newspapers at least, witchcraft and the Devil's tricks had become nonsense.[11]

No one in New York in the eighteenth century was legally charged with witchcraft, and yet New Yorkers still occasionally believed each other capable of it. An acquaintance of William Johnson told him in 1746 that he had recently witnessed "a Cart Loade of old Dutch people" hauled before a rural justice of the peace, with one accusing another of being the Devil. And William Hooker Smith, a farmer born in 1740, privately recorded his own experiences with witchcraft in the Hudson Valley in his diary of "Remarkable acurances." Smith knew of a witch named "Martha Chub, a noted hag" who led a gang of witches near Sing Sing, and recorded the claim of the ailing wife of one Thomas Bishop of that same town that she was tormented by the specters of two neighboring women, as Mrs. Howells had been. Despite the fact that Mr. Bishop "used many charms to keep the tormentors out of the house," his wife died. While, then, newspapers conveyed certainty that belief in witchcraft and the Devil was superstitious, ordinary New Yorkers were evidently not uniformly convinced. And even as newspapers printed stories mocking belief in witchcraft and magic, they still occasionally printed straightforward reports of ghosts and apparitions, as when Richard Davenport's neighbor had asserted that the hurtling stones were an example of "the Preter-naturals." So Johnson's hope in his 1729 essays that New Yorkers be freed from such "Delusions which interrupt their Happiness" did not entirely come to pass.[12]

In eighteenth-century Britain the pinnacle of the exercise of reason in developing the knowledge of the natural world was formal natural philosophy, a discipline that subsumed realms of the study of the natural world that we now allot among astronomy, physics, chemistry, and biology. If philosophy is concerned with truth, natural philosophy had truth about the natural world as its subject of inquiry. All inquiry that was directed at discovering the fundamental nature of natural phenomena and discerning their patterns and connections was encompassed by natural philosophy. And in its British incarnation specifically, natural philosophy was always believed to be a route to the deeper understanding of God's nature. Far from emptying the natural world of its spiritual significance, natural philosophers affirmed the natural world as a text of God. It was the God they read of there that was at the heart of the controversies they generated.[13]

The origins of this version of natural philosophy in Britain lay in the conflicts of the seventeenth century. The polarization of Protestantism between those who embraced the verities of ritual and clerical authority and those who embraced Calvinism generated not only a splintering of churches within Britain but also an outpouring of radical religious beliefs, beliefs that intensified the political turmoil of the period. The surge in radical religion was accompanied by a welter of doctrines concerning the natural world, doctrines that propounded occult forces, secret knowledge, and methods for foretelling the future. The decorum of natural history and the social relations that were supposed to flow from the careful comparison of plain description stemmed from efforts to find areas of agreement free from these contentions. By the early eighteenth century, many latitudinarian Anglicans in particular embraced a natural philosophy that described a harmonious and lawful universe as a buttress for their arguments that the Church of England and the Crown were similar models of lawfulness. Their hope was that a harmonious society would be the outcome of people's simultaneous embrace of a lawful universe, Church, and Crown.[14]

Few New Yorkers felt such pressure to find relief from civil and religious strife in natural philosophy. When the English had fought their Civil War, New York had been New Netherland. New York did experience civil strife when Jacob Leisler briefly claimed control of the government in the late seventeenth century, and as we saw in chapter 3, that conflict had its religious resonances. But the conflict had its origins in other realms, too. As the seventeenth century became the eighteenth, New Yorkers settled into their own chronically contentious politics, one in which armed conflict was rare and localized. Despite occasional violence, like that associated with the Stamp Act riots of 1765 and the tenant uprisings in the Hudson Valley in 1766, political conflict did not threaten civil war. And political fractures did not consistently follow religious differences, even though religious commitments among New Yorkers grew stronger in the eighteenth century. New immigrants sustained their identities through their religious commitments, as did longer-established ethnic groups. While there were occasional heightened moments of conflict—with Governor Cornbury over the establishment of the Anglican Church between 1702 and 1708 or the Anglican Bishop controversy in 1769—these weren't sustained enough to threaten civil order. Dutch churches split into factions over the distribution of theological authority in the 1720s, and Presbyterians into Old Light and New Light during the Great Awakening. But by and large, while intra-church controversies were sometimes fierce, New Yorkers generally muddled along with their variegated religious commitments, tolerating, if not exactly embracing, religious differences. Without the deep anxiety over political and theological fractures that characterized Britain, then, formal natural philosophy in and of itself generated only mild interest in New York.

Some people, however, were drawn to formal natural philosophy as it circulated ever more widely in the Atlantic world and as it came to signify a "reasonable" apprehension of the natural world, for it also became a route to fame for men of middling social origins. Benjamin Franklin came to epitomize the fame that accrued to those who "improved," but the embodiment of the natural philosopher was Sir Isaac Newton, the son of a well-to-do farmer. Newton's *Opticks*, first published in 1704, was a model of the systematic investigation of light, and his *Principia* of 1687 was the epitome of a reasoned approach to celestial mechanics. In the *Principia* Newton demonstrated that celestial bodies maintained their orbital paths through gravitational force, thus revealing a hitherto unknown natural phenomenon, demonstrating it mathematically, and enabling improved astronomical predictions. By the 1720s Newton had become a British icon. Newtonian natural philosophy became the mainstay of moderate Anglican assertions of the reasonableness of the natural world, and Newton's genius became a symbol of British intellectual superiority. As Alexander Pope wrote,

Nature and Nature's Laws lay hid in night:
God said, Let Newton be! And all was light.

Like people everywhere in the British Atlantic world, New Yorkers were made aware of Newton's accomplishments. Samuel Johnson marveled in his newspaper essays at the "wonderful Discoveries Sir Isaac Newton has made." Newspapers reported on the erection of statues in his honor in Britain. William Livingston, in his poem "Philosophic Solitude," declaimed

"IMMORTAL NEWTON! whose illustrious name
Will shine on records of eternal fame.

In 1749 the printer John Zenger politely advertised for the return the works of Newton that someone had apparently filched from his shop. The awareness of Newton was, of course, somewhat uneven. In 1744 the Maryland physician and gentleman Alexander Hamilton, on a tour of northern colonies, found an Albany doctor unworthy of his conversation when he found him to be ignorant of "the illustrious Newton."[15]

If Newton was a symbol of British genius and Britons' ability to penetrate to the innermost workings of the natural world, he had not, in the opinions of natural philosophers, explained everything. While his demonstration of the existence of gravitational force was generally acknowledged as persuasive, exactly how one body influenced another across the vastness of space was a mystery. The middle third of the eighteenth century saw an outpouring of speculations on gravity's mechanism, and among the speculators was our Joseph Morgan. While Morgan was unaware of Newton for most of his life, by the time he was in his sixties he had read enough about him to feel able to offer "Some farther Improvmt on ye Astronomical Philosophy of Sir Isaac Newton & Others" to the Royal Society. Morgan's essay, nineteen pages of crabbed writing, appeared in the middle of a longer piece he sent to the Royal Society, "The Originals of all Nations," in which he discoursed on the origins of the different races of mankind. His thoughts on Newton, however, were based on the old Aristotelian categories of matter: fire, earth, water, and air. Like the four humors familiar to colonists born, as he had been, in the seventeenth century, in classical learning these four elements were the constituents of all matter. This indicates that Morgan had not read Newton himself, for Newton described the material world as consisting simply of matter, space, and forces like gravitation. Nonetheless, Morgan seems to have acquired some familiarity with Newton's ideas. Alexander Hamilton encountered Morgan while staying in an inn in New Jersey, and conceded that "he seemed to me to talk not so much out of the way" when he discoursed on the causes of tides, the shape and dimension of the earth, and the laws of gravitation.[16]

$\mathcal{T}$he most ambitious effort to engage in formal natural philosophy among New Yorkers was, to no reader's surprise, by Cadwallader Colden. Colden had read some of this new natural philosophy at the University of Edinburgh, from which he received a degree in 1705, and he subsequently spent five years in a London abuzz with talk of it. Like the natural history of New York for which he had some ambitions, early on he apparently also had ambitions to follow in Newton's footsteps. His experiences with botany in opening up access to the learned world seem to have rekindled these ambitions. "I had pleas'd my self with the conceit of my being able to explain the Cause of Gravitation," he wrote Peter Collinson in 1745, "a point which has hitherto puzled the ablest Philosophers." The next winter he published a short treatise in New York laying out his ideas, *An Explication of the First Causes of Action in Matter and the Causes of Gravitation.*[17]

Colden began his treatise in a manner familiar to eighteenth-century readers of natural philosophy, with definitions of the material world. It was, he thought, composed of three fundamental substances: resisting matter, acting matter, and aether. Resisting matter was any substance that had extension and impenetrability. Acting matter was simply a synonym for light, matter that moved spontaneously outward in all directions. And aether was a substance undetectable by the senses that took up all space in the universe not filled by the other two. Colden then described how he imagined the three species of matter interacted such that spheres of resisting matter moved in elliptical orbits around a fixed point.

The three species of matter Colden proposed were not original. Rather, his originality consisted in what he proposed to distinguish them. In contrast to Descartes, whose fundamental distinction was between *res extensa* and *res cogitans* (matter and spirit), or to Newton, for whom the universe consisted of inert matter, forces of unknown origin, and infinite empty space, Colden's fundamental distinctions were based on the nature of power. Each kind of matter had a distinct power intrinsic to it. Resisting matter was roughly equivalent to Newton's matter, having the power to resist change in its motion (although Colden insisted that such resistance was active, the constant exertion of force). Acting matter, or light, demonstrated the power of moving, continuously and ceaselessly, unless impeded. And aether's power was to transmit across space whatever other power, whether resisting or moving, impinged upon it. Given the tumultuousness of his political experiences in New York, perhaps it is unsurprising that when Colden meditated on the nature of the world, he was so concerned with finding the origin of orderliness.[18]

Colden indeed later asserted that since the "divine Gouvenor of the Universe" was the model for governors of men, the recognition that the complex movements of the universe were based upon "a few and simple principles" should lead men to recognize their analogues in government. But Colden was attempting to do far more than simply work out in the material world what was so difficult in the political world. Natural philosophy was not merely concerned with astronomy and celestial mechanics, but was a project for understanding all of nature. Colden believed that, while he himself had reasoned from his underlying principles to celestial motion, others could begin with his principles and reason to the causes of electricity, fermentation, acids and alkalis, the circulation of blood—in short, all the phenomena of the universe. Colden could make this claim because a crucial syllogism underlay eighteenth-century British natural philosophy: God was perfection, perfection was simplicity, and therefore nature, for all its abundant variation, was simple. Natural philosophers sought the fundamental unity underlying the seeming plethora of natural phenomena, and it was this that Colden thought he had found.[19]

If Colden had simply described his three species of matter and hypothesized their interaction, few would have had any interest in his treatise. In his 1746 publication however, Colden claimed he could do much more. From his concept of matter, he said, he had derived simple mathematical methods for calculating the motion of the planets. Part of the usefulness of Newton's work had been his mathematical methods for calculating celestial motion, the calculus. But the calculus was a difficult mathematics to master and Colden's promise of a simpler mathematics was consonant with the premise of a fundamentally simple natural world. Moreover, the difficulties of precise celestial measurements and the necessary complexity of the mathematics meant that some calculations were subject to dispute. Colden's friend James Alexander, the man whose efforts to secure the New York–New Jersey boundary had been derailed by claims that the calculations of latitude had been inaccurate, was particularly enthusiastic. If Colden could make such astronomical calculations certain, Alexander wrote him, "they will be

Lasting obligations on the Learned World and transmitt your name to poster-
ity with more honour than the Conquest of Kingdoms."[20]

Alexander was not alone in encouraging Colden to expand his 1746
pamphlet. Samuel Johnson, Benjamin Franklin, and John Rutherford, a
recently arrived British military officer with intellectual interests akin to
Colden's, were all prepared to admire his efforts. Colden also sent copies
to Peter Collinson in London, who gave some to "our Greatest People in
these Studies," as well as one to the Royal Society. Although Franklin wrote
Colden that he and seven or eight other men tried reading the treatise but
couldn't understand it, and Thomas Clapp, the president of Yale, sent word
through Johnson that Colden's theory made no sense to him either, Collinson wrote of an encouraging reception in Britain. Interest in Colden's
theory was sufficiently strong that the pamphlet was pirated by a London
printer, and in 1748 Collinson wrote that it was being translated into German. In 1749, John Betts, an English astronomer, wrote Colden of his eagerness to see Colden fulfill his promise of a simplified mathematics of
planetary motion.[21]

Eager to comply, Colden wrote a greatly expanded version of his theory. The reason people had difficulty understanding his first efforts, he wrote
Samuel Johnson, "must either be in the uncommonness and newness of the
Ideas or in my inability to explain my notions properly." Consequently he
rewrote the first two chapters, adding more explanation and illustration. He
ordered the newest books on Newton from London and studied a series of
celestial observations made by the astronomer John Flamsteed. He wrote six
new chapters, chapters on the orbits and rotations of celestial bodies and on
the congruencies between what Flamsteed had observed and what his theory
would have predicted. He admitted to Betts that he hadn't quite solved at
least one problem in astronomy, but was confident that others could use his
theory to find an easy solution. He sent it all to Collinson who had it published in London in 1752 under the grand title of *The Principles of Action in
Matter, the Gravitation of Bodies, and the Motion of the Planets, explained
from those Principles*, Colden's own *Principia*.[22]

Had he produced such a mathematics it would have indeed ensured
his fame. Colden repeatedly asserted that planetary motions could be calculated using the arithmetic and trigonometry that every gentleman's son
learned at school. But what Colden actually did was to set down arithmeti-
cal progressions, adding sequential squares of numbers to get a series. He
then performed symmetrical operations, like adding and subtracting, and got
symmetrical results. The numbers in his treatise were not a measure of force

or distance, for Colden never measured anything himself. He never got up at night to measure the progress of a comet nor ordered precise instruments so he could carefully observe planetary motion. Instead, Colden used mathematics to illustrate his assumption that the universe behaved in simple and symmetrical ways. He claimed Flamsteed's observations were the "experience" that proved his theory, but never explained how he had arrived at figures that matched Flamsteed's celestial measurements. The mathematics in his treatise were not proofs but similes.[23]

Colden was sincere in his belief that the mathematics with which he illustrated his theory of matter really did demonstrate the fundamental simplicity of the universe, a simplicity that was reflected in real celestial motions. His doctrine, therefore, was both metaphysically sound and could solve concrete material problems like predicting the motion of the moon. But even though Colden claimed that he could reveal how gravitation worked and could demonstrate a simpler mathematics of celestial motion, those in his audience who were mathematicians and astronomers were not deceived. The eminent Swiss mathematician Leonhard Euler began kindly enough. "The Book," Euler wrote, "contains many Ingenious Reflections upon that Subject for a Man that has not entirely devoted Himself to the Study of it." But, Euler continued, Colden's reasoning was "destitute of all Foundation" and showed "but little knowledge of the principles of Motion." Colden got similarly incisive assessments from others in Europe, and eventually even a copy of the German translation of his first little treatise found its way to New York, with a commentary appended. The commentary was in German, which Colden couldn't read, but concluded with a quote from Ovid, which he could. "Magnis tamen excidit ausis," the quote read, "Though proved too weak, he greatly daring died."[24]

While the inadequacies of Colden's mathematics determined the treatise's reception by mathematicians and astronomers, his mathematics did not interest most of his readers. Few readers, in any case, knew enough mathematics to evaluate his treatise on that score. The *London Magazine,* for example, reviewed the treatise respectfully and declared that it was "wrote in an analytic Method, as all such Books ought to be." Others were more skeptical, with one critic sniffing, "It would be remarkable . . . if that, which was incomprehensible to a Newtone should be cleared up by a Countryman from the New World." These readers read Colden's treatise in the context of debates within the broad arena of natural philosophy, and if mathematicians and astronomers were clear that Colden had failed to contribute something useful to their field, here Colden's reception was clearly mixed.[25]

When the 1746 pamphlet was published in New York, Samuel Johnson responded with alarm to Colden's description of power as inhering in matter itself. Allowing action to arise from matter, worried Johnson, verged dangerously on atheism. If natural philosophy searched for truths about the natural world, and the natural world was a text of God, the truths discovered by natural philosophers must, everyone agreed, point back to him. An English correspondent similarly worried about Colden investing light with inherent motion, asking, "into what unfathomable Depths, both in Philosophy and Divinity, must such a notion lead?" Colden reassured his friend Johnson that he conceived of no such atheistic autonomy for matter, for matter simply acted without consciousness or intention. Colden repeated this assurance in his extended *Principles* and, for good measure, added an entire chapter on the "Intelligent Being" and the material world as evidence of his design.[26]

When Colden's *Principles* was published in Britain, it elicited little anxiety around atheism. Rather, it was embraced by people who were Hutchinsonians. Spurred by the work of John Hutchinson, who wrote a rebuttal to Newton entitled *Moses's Principia* in 1724, Hutchinsonians were Tory Anglicans appalled by natural philosophers' predilections for reading the word of God in the book of nature. Hutchinsonians claimed the reverse, that one could learn every detail about the natural world by reading the Bible in the original Hebrew. Hutchinson asserted that the Hebrew Bible revealed that air, light, and fire were three modifications of aether, thereby demonstrating in the natural world a homologue of the Trinity, a position that buttressed Tory Anglicans against the dominant Latitudinarians.[27]

The warmest embrace of Colden came from one Samuel Pike, an Englishman who, like Hutchinson, sought to know the natural world through searching the Bible. Pike had evidently come across the short version of Colden's treatise and, enthralled by what he saw as the correspondence between Colden's idea of aether and the biblical "firmament," devoted twenty pages of his own *Philosophiae Sacrae: or, Principles of Natural Philosophy Extracted from Divine Revelation* to Colden. He sent Colden a copy accompanied by an admiring letter, and Colden sent him a courteous reply, although he confided to Franklin that "his book has not increased my vanity much." While Colden repeatedly asserted that his theory in no way contradicted revealed religion, he also objected that appeals to God's omnipotence would render philosophical inquiry undeniably "very short." If everything one needed to know about the natural world was found in the Bible, then there would be no need for Colden.[28]

As with his botany, Colden's natural philosophy found him little companionship in New York itself. Aside from his friends James Alexander and Samuel Johnson, his efforts in fact drew not admiration but mockery. As the reader will recall, the late 1740s was also the time in which Colden resumed an active political role in New York, supporting Governor Clinton during King George's War. In November 1747 someone placed an "advertisement" in the *New-York Evening Post*:

> In a short Time will be Published a New Method of Controversy or an easy Way of Shortening Debates, by allowing only ONE SIDE to Publish Their Thoughts. Containing also Reasons for abolishing the Liberty of the Press.
>   By that renow'd profound Adept in the Occult Sciences
>   C———r C———n

The mock advertisement elicited letters castigating this "profound Conjuror." By calling Colden an "Adept in the Occult Sciences" and a "Conjurer," New Yorkers took note, derisively, of Colden's efforts to penetrate the secret of gravity. When Colden published an anonymous piece in January 1748 that urged the election of candidates supporting the governor, a pamphlet purportedly from "a farmer," it unleashed responses as vituperative as any that had yet appeared in what was not a very decorous press. Rather than continuing "bedawbing the weekly News Papers with the Excrements of his Brain," one letter writer advised, Colden should return to his "Starr gazing, and indulge thine Excess of Folly and Vanity . . . in thy Conceit of thy self-sufficiency for making further Improvements upon Sir Isaac Newton's Philosophy. Then," the writer sneered, "wilt thou be out of the way of doing further Mischief, or Harm, to any one but thy self."[29]

After he retreated to his farm once again, Colden got the dismaying reactions from mathematicians and natural philosophers in Europe. Convinced that his theory was merely misunderstood, he revised it, trying to satisfy his critics. But no one was interested any more. Collinson wrote him regretfully that the London printer had lost so much money on the treatise that he wouldn't print a second edition unless Colden paid for it himself. But this was beyond Colden's means. Finally he sent it to Robert Whytt, a professor at Edinburgh whose friendship he had gained through Jane's botany, who promised to put the revisions in the university's library. And there it remains to this day.[30]

*E*ven though Colden's endeavors did little to enhance the appeal of natural philosophy locally, the link forged in Britain between natural philosophy and reason influenced at least some New Yorkers to demonstrate their mastery of reason by embracing natural philosophy. Despite contentions over its founding, King's College managed to open in New York City in 1754 with Samuel Johnson as its president. Natural philosophy was part of the curriculum, taught first by Daniel Treadwell, a young man educated under John Winthrop at Harvard. He was followed by Robert Harpur, a graduate of the University of Glasgow. The dozen or so New Yorkers admitted every year, then, could expect to emerge with at least some degree of familiarity with the natural philosophy taught to similarly situated young men around the rim of the British Atlantic. Mathematical schools that taught the skills needed by navigators and surveyors began to offer instruction in Newtonian mechanics, and by 1760 the instrument seller Anthony Lamb advertised that customers could find him "at the Sign of Newton's Head." By the 1770s itinerant lecturers offered courses in natural philosophy for the edification of ladies and gentlemen, like those offered by electrical demonstrators. Even an occasional almanac maker adopted the persona of a sage and erudite natural philosopher. "Abraham Weatherwise, Gent.," for example, claimed that after extensive education and travel, he had settled down on a modest farm on Long Island, amidst "My Globes, Glasses and other Mathematical Instruments." There his neighbors constantly sought his advice, awed by his wisdom and insight, claimed Mr. Weatherwise, although, as he published only one almanac in New York, apparently the almanac-buying public in New York was less impressed. Nonetheless, by the 1760s and 1770s natural philosophy was broadly familiar to wider circles of New Yorkers.[31]

The example of the transit of Venus in 1769 demonstrates how middling men might use natural philosophy to perform their reasonableness publicly. Transits are passages of the inner planets, Mercury and Venus, across the face of the sun that are visible from Earth. Reasonably rare, they were opportunities to calculate the dimensions of the solar system. The path of the planet across the sun's disc can be thought of as one side of a triangle whose other two arms extend from the observer on Earth to the two points on the sun's rim where the planet enters and exits. If multiple observers scattered over Earth's surface all precisely measure the passage of the planet's entrance and exit in local time, and if one knows the distance between the various observers, calculations of all the resulting triangles yield the length of the arms reaching from observers to the sun. Because Mercury is relatively close to the sun, observations of its transits, of which there was one in 1753, are generally useful merely for determining the longitudes of various points on Earth. Transits of Venus, by contrast, were the transits by which astronomers could come to know the true dimensions of the solar system. Because of the necessity of a number of widely spaced, skilled observers, observing the transits of Venus became one of the great collaborative projects of the eighteenth century. The transit in 1761 was relatively little observed, as it occurred during the Seven Years' War. But the transit of Venus in 1769 occurred in peacetime and observers from all the learned nations participated.[32]

In New York, only James Alexander, who had a considerable interest in astronomy, concerned himself with the 1753 transit of Mercury. He tried to get Colden to participate, but Colden was fondest of the astronomy that went on in his own head. Alexander also tried to drum up other people's interest by writing a pamphlet on it that Franklin published in Philadelphia. Alexander with his son and another young man set up their instruments, but the day disappointed, as it was too cloudy to see anything. The efforts of John Winthrop in Massachusetts and of Mason and Dixon in southern Africa to observe the 1761 transit of Venus were reported briefly in the New York press. Then, in keeping with the widespread attention throughout Europe and its colonies, the 1769 transit of Venus merited not only one set of observations in New York but two.[33]

A group of "gentlemen" under Robert Harpur, the natural philosophy professor at King's College, made one set of observations, and the other was performed by Samuel Spencer Skinner, an English Quaker who had arrived in New York a decade earlier and set up as a rum distiller. John Holt's *New-York Journal* printed a drawing depicting the transit, which Skinner had submitted, so that New Yorkers could visualize what the observers had seen, accompanied by information like the comparative latitudes and longitudes of Boston, New York, and Philadelphia so readers could presumably compare the observations from those places when they were published (illustration 6). Engravings of any kind, other than a masthead, were rare in New York newspapers, and the inclusion of Skinner's representation of the transit indicates how significant Holt thought the endeavor was.[34]

Puzzlingly, though, neither Harpur's nor Skinner's observations seem to have been shared with anyone other than New York newspaper readers. Since the observations were useless unless the different observations from different locations were compared, the failure of their work to circulate beyond New York defeated the purpose of their endeavors. It's hard to understand why someone like Colden, who while elderly by the 1760s was still the colony's lieutenant governor, didn't forward them to the Royal

Mr. HOLT,—SIR,

BY giving the following OBSERVATIONS and DESCRIPTIVE SCHEME of the TRANSIT of VENUS, *June* 3d, 1769, a Place in your next Journal, you will oblige,
* *New-York*, near *King's-College*,
June 6, 1769.
SIR, your humble Servant,
S. SP. SKINNER.

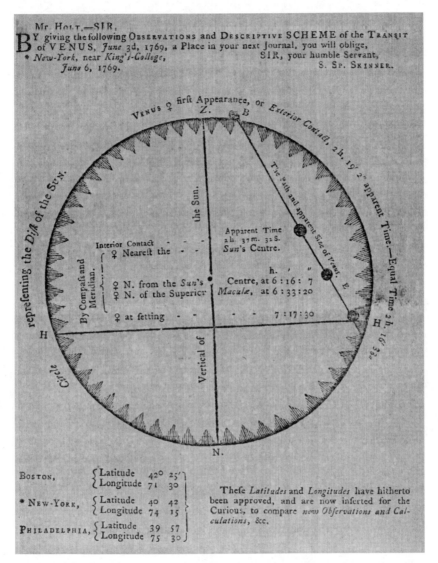

BOSTON,	{ Latitude	42°	25'
	{ Longitude	71	30
* NEW-YORK,	{ Latitude	40	42
	{ Longitude	74	15
PHILADELPHIA,	{ Latitude	39	57
	{ Longitude	75	30

Thefe *Latitudes* and *Longitudes* have hitherto been approved, and are now inferted for the Curious, to compare *new Obfervations and Calculations*, &c.

ILLUSTRATION 6. "The Transit of Venus," Samuel Spencer Skinner, *New-York Journal*, 29 June 1769. Skinner's depiction of this celestial phenomenon was intended to help newspaper readers understand the significance of the observations he described. *(Collections of the New-York Historical Society.)*

Society, or why one of the many merchants who did business in Philadelphia or Boston, both of which had groups observing the transit, didn't forward them to a connection there. Perhaps the absence of such dispatches merely affirms Colden's marginality among many New Yorkers, and suggests the indifference to such matters among many others. In any case, while Harpur and Skinner certainly performed during one of the key moments of eighteenth century natural philosophy, they seemed to have only performed it for themselves.[35]

*I*n both Colden's efforts and the attention paid to the transit of Venus one can see the centrality of astronomy to natural philosophy. While subjects like mechanics and optics were certainly part of natural philosophy, astronomy as both a subject and as a metaphor for desirable political, social, and religious relationships dominated, as was evident in the activities with which New Yorkers concerned themselves. Indeed, the revolution in astronomy begun by Nicholas Copernicus in the early sixteenth century in which the Ptolemaic model of the cosmos was replaced by a model in which the sun was fixed and Earth and other planets circled around it has often served as a synecdoche for the origins of modern science as a whole. Copernicus published his system in 1543 and sixteenth-century astronomers like Johannes Kepler and Tycho Brahe made astronomical observations that confirmed it. Despite occasional difficulties, like those of Galileo, by the early seventeenth century the Copernican system was well accepted by astronomers in Europe. Newton's discovery of gravitational force as the underlying explanation for why planetary orbits were elliptical was the capstone of this effort to penetrate to a true understanding of the natural world.

But many colonists who migrated to North America in the seventeenth century were unaware of Copernicus's innovation and took the Ptolemaic model as a matter of course. And most colonists right up to the end of the colonial period got their cosmology not from reading Newton but from almanacs. When almanacs began to be printed in New York in the 1690s, their author, Daniel Leeds, presented an Earth-centered universe to his readers. Slowly the depiction of the cosmos in almanacs changed. John Clapp's rival almanac in 1697 described the sun as "the Eye and Soul of the World" around which the planets traveled. But Clapp was the almanac maker who got the dates of the eclipses wrong, which hardly would have persuaded New Yorkers to rely on him instead of Leeds. In 1723 one "B.A." offered a complete description of the Copernican system in an almanac for New Yorkers. He wrote it, he said, "for the unlearned, that they may know the general received Opinion of the Learned World." He, like Skinner after him, included a graphic representation so that readers could visualize what the sun-centered cosmos was really like (illustration 7). Because readers would have been familiar with the symbols for the planets and the moon from their scrutiny of the Man of Signs and the monthly calendar, B.A. could use them in his depiction, confident that his readers would understand. Like Clapp, B.A. apparently published only one almanac, so New Yorkers apparently did not buy it, in both senses of the word. Leeds's son and successor, Titan, confidently published Ptolemaic almanacs at first, but by the 1730s had become uneasily aware that he was out of step with almanac makers elsewhere. In his 1734 almanac he presented the Ptolemaic cosmos in his May verse ("But by the Sun's Revolving thro' the Skies/The Needful Seasons of the Year arise") and the Copernican in July:

> Copernicus, who rightly did condemn
> The oldest System, formed a wiser Scheme,
> In which he leaves the Sun at Rest, and rolls
> The Orb terrestial on its proper Poles.

By the 1750s almanac makers referred solely to the Copernican system. The shift, then, in the cosmology presented to New Yorkers in their almanacs was reasonably slow, extending over decades, but steadily moved to reflect the astronomical convictions of the learned.[36]

Nonetheless, in 1767 Frank Freeman felt compelled to lay out Copernicanism fully before his almanac audience in order "to root out Error and Superstition." The error and superstition that so exercised Freeman was not so much Ptolemaic cosmology itself but the accompanying belief in astrology. People's convictions that as the planets shifted overhead, they shifted the relative temperature and moisture of all terrestrial matter was based, he thought, on misconceptions about the structure of the universe. The new cosmology not only shifted planetary orbits into circling around the sun, but also, as the transit of Venus project showed, vastly expanded space. What Freeman made clear to New Yorkers was that the new cosmology sundered the relationship between the motions of planets and terrestrial events. However much other almanac makers had come to accept Copernicanism, they all continued to print the Man of Signs, to calculate the movement of the moon through the zodiac, and to predict the weather. "[T]hese things," wrote Freeman, "are highly ridiculous . . . The Influence of the Planets is altogether imaginary." The best way to judge the weather, he declared, is "to wait until it comes." While Freeman's almanac had the calendar and useful tables of

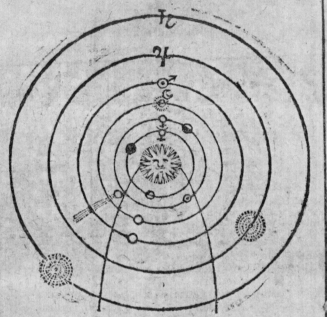

THE *Copernican* Syftem being Univerfally received and allowed by all Aftronomers, as True, I fhall give you a Reprefentation of the fame, which here follows, *viz.*

By this Figure you perceive the Sun is the Center and the Planets or heavenly Bodies (among which this our Globe of Earth is one) move round the Sun in their proper Orbs, and according to their feveral Periods.

Firft, the Globe or Planet *Mercury* is the neareft to the Sun and moves round it in 88 Days. This Planet is in Diameter 4248 Miles; in Surface it contains 55 Millions of Square Miles, and is in Solidity 39 Millions of Millions of Cubical Miles; and is 32 Millions of Miles diftant from the Sun; it enjoys about fix Times as much Light and Heat from the Sun as this our Globe of Earth. And although this Globe or Planet Mercury be of fo great

ILLUSTRATION 7. "The Copernican System," in B.A., *An Astronomical Diary*, or *An Almanack for the Year of Christ, 1723* (New York and Philadelphia: William and Andrew Bradford, n.d.). The first appearance of this sun-centered understanding of the structure of the universe in New York, this representation retains the zodiac signs for the five planets besides Earth familiar to almanac readers. *(Reproduced with permission of the Library of Congress.)*

things like coin exchange rates and postal rates, it was a baldly reasonable almanac, bare of all he deemed superstitious.[37]

Subversion of connections between planetary motion and the weather also came from another direction, from efforts to quantify the weather. For four months in 1767 "Americanior" published in local newspapers charts of temperature readings that he (presumably "he") had taken along with the accompanying weather. Asserting that "one Morsel of Fact being equal to a whole Hetatomb of Fable," Americanior offered the accumulation of such information as one component of the rational search for improvement. Such discoveries as the patterns of temperatures would help one understand "the Opulence of the Body Collective" and the rise and fall of nations. "Rhodensius" from Rhode Island wrote to the *New-York Journal* that he was pleased to see Americanior's investigations into New-York's climate, and speculated that it may well be that the region was akin to the Chinese maritime climate, both regions being on the eastern shores of continents. These measurements of local weather, then, were also part of the broad project of natural philosophy, to discover the underlying reasons for natural phenomena through systematic investigations. While it is impossible to know who among New Yorkers found Americanior's charts of interest, his work and the occasional reports of similar investigations abstracted weather from planetary influences altogether.[38]

New Yorkers, however, did not change their minds so easily, and certainly not because one man declared their everyday practices foolish. "What is an Almanac without the Weather, and the Signs?" protested one letter writer to a newspaper about Freeman's almanac. This New Yorker claimed he had decided to weigh Freeman's almanac against his own favorite, that of John Nathan Hutchins. So he and a neighbor took a good pendulum clock and timed the setting of the moon. On this particular night Hutchins had predicted 10:41, Freeman 9:09, and the two men saw the moon set at 10:45. Thus the letter writer undercut Freeman's reliability on grounds evident to everyone, avoiding the more difficult proof of correlating celestial motion with the weather or with crops. Clearly, many New Yorkers agreed. Although Freeman claimed he had sold three thousand copies of his first almanac, he admitted that "many refused them merely on Account of the Omission." The next year, having found someone, he said, "who pretends to be an Adept in the occult Sciences," he put the weather and the signs back in.[39]

No other almanac maker attempted to omit such information. And there is scattered evidence that at least some New Yorkers continued to rely on a

world in which the moon and the planets affected the everyday natural world.
Roger More intermittently gave the same sort of agricultural directions in his
almanacs that Daniel Leeds had given in the 1690s. Jared Elliot, a Connecticut minister who published essays urging the improvement of agriculture in
New York newspapers, described the way farmers coordinated cutting trees
and brush with the phases of the moon. Similarly, Lambert Borghard, a farmer in Kinderhook, recorded in his farm journal in the 1760s how he planted
according to the moon. On the other hand, Roger More included this verse
with his Man of Signs in his 1771 almanac:

> Should I omit to place this Figure here,
> My book would hardly sell another Year.
> *What* (quoth my Country Friend) *D'ye think I'll buy*
> *An Almanack without th'Anatomy?*
> As for its Use nor he, nor I can tell,
> However, since it pleases all so well,
> I've put it in, because my Book should sell.

So More had doubts about how much people used the planets' places in
the zodiac to geld livestock or determine where to draw their own blood
when healing. Perhaps, just as the four humors had been simplified to merely
general "humors" with only blood distinct, so perhaps did only the moon's
position and phase retain significance as New Yorkers went about their work
with plants and animals. Nevertheless, enough New Yorkers stuck to a cosmology whereby the celestial and terrestrial worlds were connected to thwart
Freeman's efforts to present a "reasonable" universe.[40]

If practices of coordinating the natural world with the heavens was
now disparaged among the learned in the Atlantic world, another astrological practice was even more disreputable: that of prediction. In Britain the
use of astrological prediction by all sides during the strife of the seventeenth
century had so damaged its reputation and had been so repressed by the
politically powerful that it generally disappeared from public view by 1700.
Like the relative indifference to natural philosophy in a colony that had been
uninvolved in England's Civil War, there was no such urgency around sup-
pressing prophecy in New York. Almanac makers into the 1750s predicted the effects of eclipses and rare planetary conjunctions, as they did the
more mundane effects of planets on the weather. Almanac makers certainly
explained eclipses as natural phenomena, just as they had explained their
weather predictions as rooted in skill observing concrete phenomena, and
sometimes included woodcuts depicting eclipses. An eclipse of the moon

when it was in Taurus in 1743, predicted William Birket, meant "Damage to witty and learned Men, Merchants and Tradesmen, dull Trading and a sickly Winter." But as with Titan Leeds and Copernicanism, they became increasingly aware that such predictions were scorned in the wider world.[41]

Hutchins, who had learned almanac making from Leeds and produced almanacs in New York from 1753 to 1774, at first resisted the denigration of astrological prophecy. He ascribed such scoffing to "our wise Zealots, whose thick Sculls can't penetrate into any Thing above the Earth, or out of the reach of their Nose." But, despite rhyming defiantly, "But Art, will be Art still/Let them say what they will," even he fell largely silent on predicting the effects of events like eclipses. Once, marveling at the British defeat of France in his 1764 almanac, Hutchins reminded his readers that Titan Leeds had in 1737 "foretold many Things, that (to me) seemed strange, but I have lived to see great Part fulfilled," indicating that, whatever his public silence, Hutchins was not convinced that astrological prophecy was an empty practice. But his was by now a rare defense of prediction. By the 1760s astrological prophecies were used henceforth only metaphorically, to comment on political matters at a remove.[42]

While New Yorkers seemed to accept the fading away of prophecy, they continued to object to the utter rending of material connections between the celestial and terrestrial worlds. Even more than the almanac-buying public had protested attempts to banish astrology did New Yorkers object to attempts to alter the theological significance of celestial events. Comets in particular had long been fraught with religious significance. Of all the wonders of the natural world, comets bore the least ambiguous messages. A presage of disaster, they were warnings from a wrathful God to repent and reform. In the late seventeenth century, however, the astronomer Edmond Halley, together with Isaac Newton, asserted that comets were not adventitious objects that came hurtling towards earth unexpectedly. Rather, they were bodies that moved in elliptical orbits, obeying the same natural laws as did other celestial bodies. Halley and Newton emphatically did not intend to rob comets of their religious significance. They agreed that God used comets to call attention to the decay of moral order on earth, and they hypothesized that comets had caused particular events, like the Flood, and may well cause events in the future, like the destruction of the Last Days. But during the eighteenth century, astronomers, natural philosophers, and their public generally softened the meaning of comets, transforming them from events signifying God's wrath to ones that signified his benevolence. For them, comets became one more example of the harmonious order of God's creation, an order, moreover, that could be discerned by the eyes of reasonable men.[43]

These new interpretations of comets circulated in New York, as they did throughout the British Atlantic. During the appearance of a comet in 1744, James Alexander took careful measurements of its movements while John Peter Zenger reprinted an essay on the natural philosophy of comets in his newspaper. Jesse Parsons explained the elliptical orbits of comets in his almanac of 1757, as "B.A." had done for Copernican cosmology in 1723. The 1758 arrival of the comet whose reappearance Halley had said would affirm the theory of elliptical (and thus predicable) orbits was widely reported in New York newspapers. In June readers were informed that Halley's comet "may now be look'd for in the North East, for about 3 Hours before Sunrise: and in the North West for next four Hours after Sunset," so that New Yorkers could see proof of Halley's theory for themselves. "[L]ast Year," reported the *New-York Weekly Post-Boy* in 1760, "the learned world was gratified with the expected Visit of the Comet of 1682, which has greatly confirmed the Newtonian Philosophy, especially as to the Theory of Comets." But however widely new ideas about comets' motions circulated in New York, their religious meaning was not so easily settled.[44]

When in September of 1769 another comet appeared over New York, Samuel Skinner, the rum distiller who had published his observations of the transit of Venus the previous June, seized the opportunity to further display his mastery of astronomy. In an eight installment "Philosophical Description of Comets" printed in two New York newspapers, Skinner shared both his measurements of the comet's progress and his knowledge of much of the learned world's current thinking about comets. "Comets by all our Modern Philosophers," he informed his readers, "are said to be Opake, Compact, fixed and durable Substances." Like other natural phenomena, comets had a taxonomy. This, thought Skinner, was possibly "a Comet Crinitus, or hairy comet, throwing forth beams like hair, from every part around it." He reminded readers that like planets, their motion "always describe equal Areas in equal Times." But the public reception of this comet did not stay within such cool reasonableness.[45]

Besides Skinner's extended essay, New York newspapers printed six other reports of the comet, four letters from readers, and three poems. One of the first letters was an unsigned one from New Jersey, whose author calculated that the tail of this comet might well rake the earth. "[T]herefore," the letter writer concluded, "it becomes all to be prepared for Consequences so alarming as those which must then follow." A letter writer from Connecticut drew a similarly grave conclusion from watching the comet's path. Only heaven, this author wrote, could prevent "its dire encroachments and deserved invasions to an ungodly, treacherous, degenerate, and defenseless world." But someone else quickly submitted an alternate reading of the comet:

> The guilty nations tremble. But, above
> Those superstitious horrors that enslave
> The fond sequacious herd, to mystic faith
> And blind amazement prone, the enlightened few,
> Whose godlike minds philosophy exalts,
> The glorious stranger hail.

Like lightning, comets were an opportunity for "the enlightened few" to demonstrate their calm mastery of a now-understood natural phenomenon, in marked contrast to those in thrall to "superstitious horrors." Far from being a harbinger of disaster, the author of this poem knew the purpose of the comet was

> To work the will of all-sustaining Love:
> From his huge vapory train perhaps to shake
> Reviving moisture on the numerous orbs,
> Thro' which his long ellipsis winds; perhaps
> To lend new fuel to declining suns,
> To light up worlds, and feed the eternal fire.

While naturalized, comets nonetheless surely served God's benevolent purposes. As the poems had, a letter reprinted from the *Boston Gazette* took heated issue with the New Jersey letter, characterizing the predictions of the comet's path as "idle and contemptible."[46]

These then, were the terms of the debate: was the comet punishment due a "ungodly, treacherous, degenerate, and defenseless world" or was it a vehicle of "all-sustaining Love?" Was God a god of wrath who used the natural world to demonstrate his judgment on sinners, or was he a god of love who sustained a harmonious and orderly natural world for the support of his children? As all these contributions but Skinner's were unsigned, it is hard to know to what degree the two interpretations were related to formal religious affiliation or social position. But even Skinner was uncertain what spiritual message the comet brought. On the one hand, "many eminent Philosophers," he wrote, "have hitherto esteemed Comets as PRODIGIES, and . . . that they are OMINOUS OF FUTURE EVENTS." Skinner pointed out their historic appearance at times like Julius Caesar's murder, the destruction of Jerusalem, and the English Civil War. On the other hand, if the comet's tail did rake the earth, Skinner hoped that it "may be productive of the most salutary Consequences" through the transmission of "such well-rectified Matter as the Vapour of the Comet is capable of." Thus he tried to please all parties in his audience.[47]

$\mathcal{T}$he public contention over this comet should not surprise us. In the months before its appearance the Anglican bishop dispute broke out, agreements over boycotting British goods fell apart, the improvement society was imploding, and the contest for seats in the General Assembly was as vicious as any in the colony's history. If some New Yorkers now thought that astrology was nonsense and that astronomy demonstrated God's harmonious order, others refused to abandon their convictions that God used planetary motion to effect change on earth and comets to warn of his wrath. These New Yorkers refused to concede that their beliefs were superstitious, for they found them affirmed by their everyday experiences and by their theology. In a colony in which political, economic, and religious fractures were eminently visible, people who insisted that their familiar beliefs about the natural world were reasonable had no hesitation making these assertions public.

There is another reason why some New Yorkers resisted the new version of a reasonable natural world, one more subtle but equally important. New assertions about what constituted a reasonable understanding of the natural world—that witches and ghosts were imaginary, that astrological influences were physically impossible, and that the cosmos demonstrated regular patterns according to natural laws—were epistemologically at odds with familiar ways of knowing and thus challenged familiar social relations of knowledge. In New York as elsewhere in the early modern world, the touchstone of truth was one's own material experience, and knowledge of the natural world was within the reach of anyone who bothered to pursue it. Those who now asserted that the sun was fixed and it was the earth that moved demanded that people disbelieve the evidence of their own senses, for everyone saw for themselves the sun moving through the sky. Astronomy in particular, the realm of knowledge that was the core of what we now call the scientific revolution, asked people to believe a cosmology sundered from their own sensory experiences.

Moreover, astronomy required both instrumentation and calculations increasingly beyond the reach of most people (even Colden, certainly among the most educated men in New York, found the calculus beyond him.) Therefore, people had to rely on the assertions of others about the true nature of celestial phenomena. Unlike improvements, which people could try for themselves, or refined knowledge, for which people could, if they were so inclined, count stamens and pistils, participate in an electrical demonstration, or try a patent medicine that promised to strengthen lax fibers, few in New York (or anywhere else in the Atlantic world) could experience the truth of new assertions about the cosmos. To accept the new cosmology was to trust the assertions of those erudite few who did the observations and calculations upon which an astronomy of reason was based. Such social relations that called for trusting the distant erudite were antithetical to the familiar and satisfying ones in which people relied on themselves and those they knew to be trustworthy through face-to-face encounters. Without powerful reasons to embrace the new cosmology, many people stuck with what they could see for themselves.

By the 1770s, then, while some assertions about what understandings of the natural world were within the bounds of reason and what fell outside those bounds had had an impact in New York, others were resisted. It is tempting to analyze this as a developing split between elite and popular understandings of the natural world, and certainly a man like Samuel Spencer Skinner tried to demonstrate his gentlemanly reason through his endeavors in astronomy. But the social distribution of knowledge in New York was not so clear cut. It is probably more useful to think it as differences between people more oriented towards the Atlantic world and those more concerned with local relationships. Colden, in particular, was someone who sought esteem and friendship elsewhere to compensate for what he found so little of close to home. But Samuel Johnson, equally educated and equally sensitive to intellectual currents in the wider Atlantic, was less sure about the boundary between the reasonable and the superstitious. His son once told him of an ailing acquaintance who sought "the applications of a seventh daughter." To be the seventh child of a seventh child was, in common belief, to have special healing powers. "I have no faith in a seventh daughter or son either," Johnson wrote back. But, he added, "that is a strange story Captain Porter tells of his daughter, and if Mr. Whittelsey must try such a means try that seventh son at Derham." Perhaps the healing power of a seventh child was just superstition, but someone whom Johnson trusted had had a material experience of its efficacy, and such testimony could not be discounted. Almanac makers, all middling men located in and around the port of New York, also reflected this ambiguity. They were clearly influenced by intellectual currents circulating in the wider world, yet their largely rural readers needed the concrete guidance of weather predictions and the phases of the moon, and cared little about the musings of natural philosophers.[48]

If some abhorred magic, it was not necessarily because they now acknowledged it as foolish superstition. "An old Indian come here to day that lets fortans and ueses charmes to cure tooth ach and drive away rats," wrote Mary Cooper, the Long Island farmwoman, in her diary in 1769. "O Lord, thou knowest that my soul abhors these abominations." For her, magic was abhorrent for its association with devilish ways, not foolish because of its material impossibility. Uninterested, and probably largely unaware, of the transformations in what were reasonable understandings of the natural world, Mary Cooper nonetheless shared one near universal conviction with her fellow New Yorkers: that the natural world was a text of God, one in which one read his nature.[49]

CHAPTER 7

Landscape Reimagined

⚶⚛❀⚛⚶

Sweet Liberty, thou pleasing thought,
Which our fore-fathers to America brought,
Being then an universal howling wilderness,
Where many Savage Indians did them distress[.]
        Copernicus Weather-Guesser,
    The New-Jersey Almanack for . . . 1769

*T*hus did Copernicus Weather-Guesser
read the meaning of space through a taxonomy of race. In his verse colonists subdued both the material wilderness that was America and the wild humans who inhabited it. Armed with the "Sweet Liberty" that gave each white man his autonomy, colonists transformed the undifferentiated space of wilderness into distinct, civilized places, while emptying them of an uncivilized people. By 1769, when Weather-Guesser published his almanac, the literate-but-not-learned New Yorkers who were his readers increasingly embraced this understanding of themselves and of the place they inhabited. If in the seventeenth century colonists had seen a wilderness that they were struggling piecemeal to remake into a domesticated and productive landscape, over the course of the eighteenth century they increasingly came to think of New York as a site for grand struggles between civilization and savagery. Increasing awareness of the geography of elsewhere raised awareness of the significance of local geography on the world's stage. And "New York" became increasingly understood as a specific, bounded place, rather than a general area that shaded off gradually until it merged with the vastness of the continent. Simultaneously, literate New Yorkers recast the meaning of the human landscape of the colony. By the 1760s Weather-Guesser's readers had largely shifted from categorizing inhabitants of New York as Christian and heathen to categorizing them as white, black, or red, categories that increasingly seemed innate, natural. As they did so, they reimagined the human landscape as one consisting of the naturally civilized and the naturally savage.

In the early modern period, Europeans increasingly applied techniques we now associate with formal science to the knowledge of territorial space and to the distribution of human difference. Measuring, describing, categorizing: all were applied to understanding the topography of the earth and its natural and human variation. Cartographers assembled measurements of distance and direction, longitude and latitude, and plotted them on ever-more-accurate maps. Printers of gazetteers and geographies gathered increasingly detailed descriptions of places and what could be found there, human and nonhuman. The widespread assumption that differences in people could be attributed to the effects of their different environments, as Jaspar Dankers had assumed when describing colonists around Albany, faded from prominence. Increasingly, describers of human difference arrayed groups of people into taxonomies of "nations" or "races" that were presented as if they were inherently distinct from each other.[1]

The prestige of geography, cartography, and new constructions of human difference was enhanced by their associations with improvement, refinement, and reason. In the eighteenth century throughout Europe and its empires, geographic knowledge and cartographic expertise became emblematic of improvement, for both facilitated commercial and imperial expansion abroad and prosperity at home. Geography became a realm of refinement, an area in which the genteel, elevated above personal and local experiences, could claim to know people and places they themselves had never experienced. New understandings of differences among humans were similar to botanical taxonomies, systematic categorizations that allowed the person of sensibility to distinguish the civilized from the rude. And through the perusal of maps and descriptions of places and the people in them, the socially elevated could presume to know both the natural and the human world as they really were through reason.

In New York, as elsewhere, the circulation of maps and of geographic descriptions rose over the course of eighteenth century, and differences among groups of people were increasingly described with racial idioms. As with other realms of knowledge of the natural world, the particular patterns of circulation and adoption reflected local social relations, particularly concerns with self-interest, both one's own and that of others. When interests were united, adoption was relatively straightforward. The kinds of maps and information useful to navigators, for example, expanded in tandem with the expansion of the commercial networks New Yorkers had been developing since the seventeenth century. Anything that improved navigation was taken for granted as in everyone's interest, and experiences with seafaring that indeed became faster and more reliable in the eighteenth century affirmed the association of maritime cartography with improvement.

Because of their interests in the labor of slaves and in the land of Indians, white New Yorkers moved fairly easily to describing Africans and Indians as belonging to races distinct from their own, even though for political reasons the Iroquois were relatively protected from this for a time. Although the adoption of such racial categorization required that white New Yorkers ignore the evidence of their own experiences, self-interest propelled them to near unanimity in embracing this categorization. But even as white New Yorkers increasingly proclaimed the natural savagery of Africans and Indians, people in the colony who were African and Indian were increasingly like other New Yorkers in their material and their spiritual practices.

New Yorkers were more ambivalent toward territorial maps and geographic descriptions than they were toward racial categorization, even though their circulation rose dramatically, especially after mid-century. Cartography in particular remained suspect, because maps produced and circulated in New York served some New Yorkers' interests to the detriment of others.' The ways in which these new configurations of the knowledge of place and race were adopted by white New Yorkers, then, were shaped in large part by the relationships between knowledge and self-interest.

The dissemination and reception of this rise in geographic knowledge and this reconfiguration of human differences in New York was uneven and related to particular local experiences. A major link between local events in the colony and New Yorkers' reimagination of their colony's geography and its human inhabitants was the rising conflict between Britain and France. Between 1689 and 1763 France and Britain, with the occasional participation of Spain, fought four wars, all of which entailed some action in New York. While the rising tide of commerce certainly stimulated the wider circulation of maps, warfare eventually had an even more powerful effect both on practices of cartography and on local geographic imagination. Because the Iroquois were crucial to British claims to rights to the interior of the continent, and because New York's claim to importance as a colony stemmed largely from its geographic position as the gateway to the Iroquois, the Iroquois emerged as distinct from other Indians. In the wake of the defeat of France in the Seven Years War, however, New Yorkers no longer needed the Iroquois as strategic allies and now lusted after their land. Abruptly the Iroquois were dissolved into a general Indian savagery, for it now served New Yorkers' interests to embrace new racial categories.

Africans and their descendents, by contrast, were difficult to incorporate into narratives of the struggle between France and Britain, or to be useful to elevating the significance of New York. In the wake of the 1712 slave rebellion, white colonists had their suspicions of their slaves affirmed by reports of conspiracies and violence elsewhere. Then an incident during the war with Spain in 1741, as we shall see, ratcheted up white New Yorkers' hostility toward black New Yorkers, and facilitated the embrace of new racial idioms by whites.

Both this war and the subsequent Seven Years War raised the visibility of North American geography among New Yorkers and the place of their colony within it, although not without dissension. By the end of the colonial period the ways in which whites in New York had reconfigured both the local human landscape and their understanding of the geographic significance of their colony reflected not only their own interests but also their experiences within a wider landscape of war and imperial politics.

*I*t was Iroquois warfare, as much as conflict with the French, that initially prodded at least some New Yorkers to attempt to raise the geographic visibility of their colony internally and in the wider Atlantic world. In the seventeenth century, the Iroquois had warred against other Indians in all directions, and through these wars became an enemy of the French in Canada. During the last third of the century, various Iroquois nations had made a series of alliance agreements, collectively called the Covenant Chain, with New York officials as a counterweight to French enmity. By 1700 the French had turned back Iroquois efforts to extend their influence northward. Consequently, in 1701 the Iroquois deeded a claim to their "beaver hunting grounds," land north of Lake Erie, to the Crown to hold for them in trust. The conveyance of land north of Lake Erie was intended to shift responsibility for land the Iroquois no longer controlled to their allies against the French. New York officials who accepted the deed ignored the Iroquois' doubtful claim to the territory, as it was in their interests to present the Iroquois' possession as fact. As had Governor Bellomont when he sent the map that Colonel Römer had drawn, New York officials pointed out to London recipients of news of the deed that access to that territory was through New York. Shifting their emphasis from the fur trade, New York officials in the eighteenth century stressed that their position abutting the Iroquois was vital to upholding British territorial interests against the French. Keeping the Iroquois in alliance with Britain via the Covenant Chain and away from relations with the French in Canada became one of the major justifications New York officials henceforth offered when arguing for attention and money from London.[2]

In 1710 some enterprising New Yorkers, including Colonel Peter Schuyler, an Albany fur trader and prominent commissioner for Indian affairs, brought four young Iroquois men with them on a trip to London. There the New Yorkers showed them off as "Indian Kings" (even though only one could plausibly be described as even a sachem), touring them through London sights, introducing them to leading officials and clergymen, and—the greatest coup of all—presenting them to Queen Anne at court. This representation of the Iroquois as ruled by noble lineages that paralleled Britain's own was intended to strengthen Britain's commitment to New York. And, by and large, to the degree London officials were interested in New York at all, they agreed that much of its importance stemmed from its geographic location adjacent to the Iroquois. In 1721 the Board of Trade, for example, compiled a monumental report on the state of the colonies in which they recognized New York's significance because of its proximity both to the Iroquois and the French.[3]

New Yorkers' argument that they had obtained rights to the interior of North America for the British crown was disputed by French officials, for they had their own interests to uphold. In 1718 Guillaume DeLisle, one of the most skillful cartographers of the early eighteenth century, engraved a map that among other things disputed New Yorkers' claims. DeLisle's "Carte de la Louisiane et du Cours du Mississippi" showed all the waterways in the interior of North America that had been explored by the French (as well as some speculative ones) laid out under a grid of longitude and latitude. Its claims to measured exactitude bolstered the political claims of the map, for this was a map of more than mere topography. It also made a particular argument about the basis for sovereignty, one that French diplomats were beginning to make regarding the relative claims of Britain and France to North America. The French claimed that the entire watershed of a traveled waterway was encompassed by the right of discovery. Accordingly, areas that were drained by rivers that ran into the Atlantic, like the Hudson, were recognized as belonging to the British. The lakes and rivers of the territory of the Iroquois, however, ran into Lake Ontario, a waterway the French had reached first. Consequently, DeLisle drew a line on his map that neatly bifurcated New York's water and portage route from the Hudson to Lake Ontario, the one Römer had followed. Thus did his map deny New York officials' main claims to importance within the British Empire, namely that the 1701 Iroquois deed had given the crown sovereign rights to territory in the interior of North America, and that access to that territory was through New York.[4]

In response, Governor Burnet and his Council had William Bradford print a map in 1724 to counter DeLisle's assertions (illustration 8). Even more clearly than with Römer's map, on this map the eye could trace the supposed ease with which people could move between Albany on the right side of the map and the upper Great Lakes on the left. Critically, the New York version put the area north of Lake Erie right in the center of the map and labeled it "The Countrys Conquer'd by the FIVE NATIONS," insisting visually that through New York lay the Iroquois claim to part of Canada, and upon this rested any British claim. Cadwallader Colden, who was a member of Burnet's Council, simultaneously set to work on *The History of the Five Indian Nations Depending on the Province of New-York in America*, which he published in 1727. The Iroquois were, according to Colden, "the Fiercest and most Formidable People in North America, and at the same time as Politick and Judicious as well can be conceiv'd." Consequently they were both necessary to colonial security and responsive to proper diplomatic overtures. While admitting that they were a "poor Barbarous People, under the darkest Ignorance," Colden also asserted that "[n]one of the greatest Roman Hero's have discovered a greater Love to their Country" than the Iroquois. They were, then, very suitable allies of the British. Colden described how important and effective New York officials' relationships with the Iroquois were in securing the Iroquois' allegiance to Britain, thus reinforcing the assertions on the 1724 map.[5]

Unlike the handful of maps produced as manuscripts in the colony in the seventeenth century, like the ones Bellomont had sent his London superiors, this map was printed. A printed map could command a wider audience, raising the visibility of the arguments on the map both in London and within New York itself. While the map certainly promoted British imperial interests against the French, New York officials also used it to oppose some other New Yorkers' interests. Indeed, many people in the colony were indifferent to these officials' claims about the significance of New York's geography.

By the 1720s, Britain and France had fought two wars. Called King William's War (1690–1697) and Queen Anne's War (1702–1713) in the colonies, both had some impact on New York. During King William's War in 1690, Schenectady was raided by a force of French and Indians, with sixty-two colonists killed, twenty-seven carried away into captivity, and the rest fleeing terrified through the snow to Albany. That summer a few New Yorkers marched in an expedition against Canada, and others marched in similar expeditions in 1709 and 1711 during Queen Anne's War. But none of these were successful, and New Yorkers resented the taxes used for

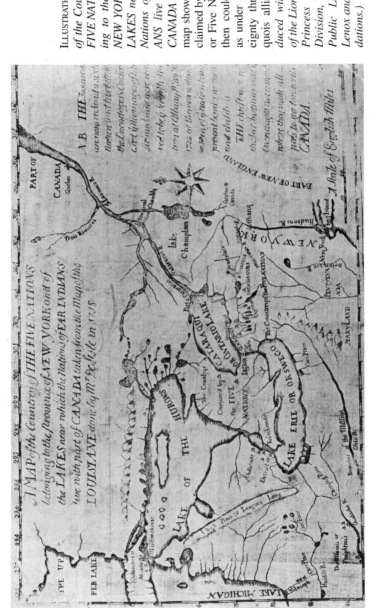

ILLUSTRATION 8. *A MAP of the Countrey of THE FIVE NATIONS belong-ing to the province of NEW YORK and of the LAKES near which the Nations of FAR INDI-ANS live with part of CANADA (1724). The map shows the territory claimed by the Iroquois, or Five Nations, which then could be claimed as under British sover-eignty through its Iro-quois alliance. (Repro-duced with permission of the Lionel Pincus and Princess Firyal Map Division, The New York Public Library, Astor, Lenox and Tilden Foun-dations.)*

frontier fortifications, which the Crown declined to fund. Moreover, many New Yorkers were indifferent to British imperial ambitions, and some were perfectly happy to do business with the French. Upon his arrival in New York in 1720, Governor Burnet had moved against Albany traders who sold British goods north to Montreal. Burnet intended to redirect New York trade westward toward the Iroquois and beyond them to the Indians of the Great Lakes. His interdiction would thus simultaneously undercut the French fur trade and promote relations with the Iroquois. The Albany traders, however, had no particular loyalty to this imperial project and had found the Montreal trade easy and profitable. By printing the 1724 map, New York officials intended to expose the disloyalty of the traders' interests and demonstrate the wisdom of Burnet's policy of interdiction.[6]

Albany traders hardly needed a map to understand where their interests lay, and luckily could take advantage of the modest understanding of North American geography elsewhere. In response to Burnet's ban, the Albany traders enlisted their London suppliers to exert pressure on their behalf. The London merchants argued that the trade was inevitable because of the geography of New York. The only route that fur-trading Indians could take to Albany, claimed the merchants to the Board of Trade, was via the St. Lawrence River and therefore they had to pass through French settlements. Anything that alienated the French would stop such trade altogether. And anyway, they continued, the Iroquois live on the St. Lawrence just across from Quebec. The French were bound to dominate their trade, so the French might as well get their trade goods from Albany as elsewhere. London merchants were not alone in their muddled understanding of New York's geography. The Board of Trade was sufficiently uncertain of the geography of the area, and took their complaints seriously. The map, printed as the frontispiece to a rebuttal of the Albany traders and their London allies, was intended to defeat them politically by exposing the absurdity of their geography. To say that the Five Nations and the Far Indians have to pass Montreal to get to Albany, wrote the Governor's Council in its rebuttal, was like saying that "one cannot go from London to Bristol but by way of Edenburgh." As Burnet pointed out in his communication to the Board of Trade, "A meer inspection of this Mapp is indeed sufficient to confute them." But whatever the hopes of New York officials to defeat their opponents through their map, the interests served by the map didn't trump the interests of the Albany traders. The London merchants prevailed, and George II nullified the ban on the Albany-Montreal trade.[7]

In this dispute, the London merchants and members of the Board of Trade certainly had only a modest grasp of North American geography. In

the eighteenth century geography books and gazetteers, increasingly produced in Britain in response to rising awareness of the value of Britain's
worldwide commerce, included descriptions of New York among their array
of the world's places. These accounts of the colony were of variable quality.
John Oldmixon's two-volume *The British Empire in America* of 1709 devoted fifteen pages to a fairly reliable account of New York's history, trade,
climate, soil, and Indians, even though at one point Oldmixon described
Long Island as being New York's northern border, an error uncorrected in
subsequent editions. Herman Moll's *The Compleat Geographer* of 1723, by
contrast, was a likely source of the misapprehension that the Iroquois lived
on the St. Lawrence (he also described their territory as stretching southwest, as far, in fact, as "Louisiana"). In Moll's account New York was a
minor place: "a narrow slip of land lying between New-England on the East,
and Hudson's River on the West."[8]

But whatever the slow rise in visibility of New York in formal geographies, interests in London and Albany trumped the weight of the argument
made by Burnet and his council that the geography of New York naturally
supported their political policies. Although London authorities consistently
stationed a garrison of British soldiers in New York and occasionally paid
for diplomatic gifts to the Iroquois, London purse strings mostly stayed shut.
And even the garrisoned soldiers were notoriously neglected by the Crown,
poorly clothed, poorly fed, and poorly supplied. Indeed, for much of the
colonial period it was difficult for New York officials to get the attention of
officials in London at all.[9]

*W*ithin New York itself, geographic information rose slowly in the
early eighteenth century. With the advent of newspapers in the 1720s, New
Yorkers now could read about events in other places. Indeed most of the news
that appeared was reprints from the London papers, with news from other
colonies and from within the colony itself growing only slowly through the
1730s and 1740s. But other than news, the geographic information offered
them in print before the middle of the eighteenth century was sporadic and
rather eclectic. In 1706, for example, Daniel Leeds offered his almanac readers "A Geographic Description of the World, Containing Europe, Asia, Africa,
and America." His brief synopsis listed various territories and countries with
their general dimensions in miles, though one wonders what his readers made
of the list's assertion that "America" consisted of Mexico and Peru. In 1728
the *New-York Gazette* offered readers a description of the Turkish empire and
in 1730 a description of trade with Guinea; in 1737 the *New-York Weekly*

Post-Boy devoted the lead article for several weeks to Confucius. Private readers sometimes read travelers' tales, as did Joseph Morgan who apparently read Hennepin's account of his travels in the interior of North America. To some modest degree, then, New Yorkers who were so inclined could acquire at least a rudimentary knowledge of the panorama of the world's places.[10]

The one clear area in which geographic information rose in the colony was in the expanded availability of maritime maps, for by the early eighteenth century New York ships sailed to all the British colonies along the Atlantic coast and in the Caribbean (and illicitly to French and Spanish islands), and to a half-dozen ports in Europe. When William Bradford opened his printer's shop in New York City in the 1690s, some of his earliest advertisements were for maritime maps and books on navigation. By 1733 he offered "several sorts of Sea-Books, Sea-Kalendars, Mariners Compass Rectified, Atkinson's *Epitome*, Newhouse's *Whole Art of Navigation*, Norwood's *Practical Navigation*, [and] the *English Pilot*." In his advertisement Bradford could abbreviate Atkinson's full title, which was *The Epitome of the Art of Navigation*, because it was so popular that he knew readers would recognize it. The *English Pilot* was a "waggoner," a collection of harbor charts and views of ports as one approached them from the sea.[11]

New York officials early understood that determining the colony's location mathematically would be a contribution to the maritime mapping so necessary to commerce. In 1679 a group of men including the colony's surveyor general, Philip Wells, and Andrew Norwood, the son of the author of *Practical Navigation* in Bradford's ad and briefly a resident in the colony, determined that the latitude of the fort on Manhattan was 40 degrees 20 minutes north. In 1723 Governor Burnet, aided by Cadwallader Colden and James Alexander, determined the longitude of that same spot. These two measurements could anchor the harbor of New York on maps that served navigators. Among the most important skills taught at the evening schools for mathematics that began to open in the 1730s was navigation. By the 1750s a half-dozen booksellers and shopkeepers in New York City offered maps and waggoners for the entire Atlantic world.[12]

But most New Yorkers didn't pilot ships or travel far, nor did travelers by land yet use maps. As people had in the seventeenth century, people traveling by land generally asked for directions or hired a guide. By 1744, when Alexander Hamilton toured the northern colonies, he could consult *The Vade Mecum for America*, a book that listed roads and taverns from Virginia to Maine. Beginning with Daniel Leeds's 1694 almanac, almanacs sometimes printed the distances between places like New York City and Albany or Philadelphia.

But almanacs and the *Vade Mecum* merely printed point-to-point lists, not maps. As such, they merely set down in print what travelers experienced as they moved from place to place. Nor did New York yet have many men with the skills to make maps, even had there been a market for them. The New York officials who had commissioned the 1724 map made no claim to new mathematical observations of natural landmarks. Rather, William Bradford simply copied the segment of DeLisle's map relevant to New Yorkers' concerns and wrote over it New York's claims to possession. And this map showed little enough of the colony itself. It was enough to locate New York in general along the right-hand edge of the map, for what was significant about New York's geography here was its relationship to someplace else. In 1731 James Lynn, the man who opened the first evening mathematics school in the colony, did engrave a map of New York City, which coincided with issuing a new charter of incorporation for the town and which was offered for sale locally. But most people most of the time could get by without any maps at all.[13]

Aside from the few engaged in disputes like that over the trade with Montreal, New Yorkers did not have much stake in revising their assessment of the human landscape of the colony. Coastal Algonquians and Indians along the Hudson River continued to live in the colony, farming, hunting, and fishing, albeit in smaller numbers. During Alexander Hamilton's travels through New York in 1744, he encountered Mohicans near Albany, saw Algonquians gathering oysters off the Bronx, and was waited on by an Indian servant girl named Phoebe on Long Island. While colonists generally still showed little interest in Indian knowledge of local plants and animals, nonetheless many Albanians were skilled canoeists, colonists on the Hudson made rope from elm bark, and the minister Henry Melchior Muhlenberg was served raccoon when he stopped off at an inn in the Catskills. An Indian knew Cadwallader Colden's interests well enough to show him twin baby raccoons as a specimen of natural history, and Albany women bought moccasins, baskets, and brooms from Mohican women. Clearly, then, Algonquians and Hudson River Indians continued to be part of the ordinary social fabric of the colony. Yet their affairs now rarely merited mention in the records of New York's governors, and complaints to local authorities, such as an incident in which Albany traders seized Mohican corn in payment for debts, seem to have often enough been brushed over. No whites had any particular interests served by revising relationships with Algonquians and Hudson River Indians, and so they drifted to the margins of visibility.[14]

By contrast, the hostility toward slaves that developed among whites in the late seventeenth century became chronic. Once newspapers began to be published in the colony, New Yorkers could read almost once a year about rebellious slaves somewhere rising up and committing "barbarous" murders, affairs much like their own experience of revolt in 1712. In the 1730s, for example, New Yorkers were offered accounts of slave rebellions in New Jersey, Virginia, South Carolina, St. John's, Antigua, and aboard at least two slave ships, as well as repeated accounts of slave-white violence in Jamaica. Occasionally local crimes committed by slaves merited reporting, as the attempted rape of a young white woman for which a slave was burned at the stake in 1734, or the slave who killed his owner and another man up the Hudson in 1737. In 1731 the General Assembly made the colony's laws applicable to slaves even harsher. And in 1737 a proposal was floated to finance cash awards to encourage hemp production through taxes on slaves, something touted as having the additional benefit of discouraging their continued importation. But whatever the evidence of risks to social peace that slavery posed, the lure of slaves' labor proved too strong, and New Yorkers continued to hold Africans in slavery and to import more.[15]

By about 1740, then, the geographic understanding of New York as a place and of the meaning of human difference within it had changed only modestly from that of colonists in the seventeenth century. The geography of the colony itself was only modestly more visible, and this mattered to ordinary colonists probably little, although maritime maps were in more demand. The Iroquois were more politically important while the Algonquians and Hudson River Indians were declining, and white attitudes toward blacks in the colony showed a sustained hostility.

Over the next twenty-five years the relationship of New Yorkers to geography would be transformed, driven in part by wide-scale changes like expanding commerce and war, and in part by events peculiar to New York itself. The circulation of maps and descriptions of other places would rise dramatically, and, in parallel, more and more detailed maps of New York would be produced. Readings of the significance of New York on the world's stage would expand and change. As importantly, beliefs and practices about who were genuinely New Yorkers would sharpen, with more hostile and adamant exclusion of Indians and Africans. The explanation for such exclusions would change as well, even as, ironically, both Indians and Africans in and near the colony become more like white colonists.

The first event that led New Yorkers to reimagine their place in the world began with a minor robbery in late February 1741. Thieves took bolts of cloth, silver items, and coins from the house of Robert Hogg, a New York

City merchant. City officials, following up a tip given by a sailor from a British man-of-war, went to the house of one John Hughson, a man who ran a public house that catered to the lower sort—sailors, servants, slaves—and who was already suspected of being a fence for stolen goods. They found nothing, but Hughson's sixteen-year-old indentured servant, Mary Burton, confided to a friend that Hughson and his wife Sarah indeed regularly received stolen goods and hid them cleverly. The friend passed Burton's accusations on, and the sheriff returned to Hughson's to find the suspect goods. John and Sarah Hughson, along with a woman who lived with them, Margaret (Peggy) Kerry, were brought up on charges, along with two men, Caesar and Prince, both of whom were slaves and who allegedly had carried out the robberies.[16]

As the investigation of these five got under way a fire broke out at Fort George on the tip of Manhattan, which burned the barracks and the governor's house to the ground before it was controlled. The next week there was another fire, and the week after that, three more. Between the first fire at the fort on March 18 and April 6, a total of ten fires were set by a person or persons unknown. Mary Burton, testifying in the robbery case against the Hughsons, Kerry, Caesar, and Prince, now let something astonishing slip: just as in 1712, these fires had been set as part of a conspiracy to burn the town, kill many of the inhabitants, and seize their possessions. Between May and August 1741, accusations of involvement in the conspiracy proliferated. By the end, twenty whites had stood trial (three others skipped town), as had one hundred thirteen slaves (with seven more accused slaves successfully disappearing.) Of the whites, four were convicted and hung, six others were banished from New York, and another found it worth his while to relocate. Thirty-one slaves were executed, and seventy others were transported out of the British Empire altogether.

One of the judges at the trials, Daniel Horsmanden, published an account of the testimony along with his interpretation of the origins of the conspiracy. The judges and jurymen, Horsmanden wrote, found that this was no mere local uprising. This supposed conspiracy took place against the backdrop of a war, the so-called War of Jenkins's Ear (named for a British commander whose ear had been cut off by the Spanish coast guard) that had begun between Britain and Spain in 1739. One of the whites in the supposed cabal, John Ury, magistrates found, was secretly a Roman Catholic priest (although Ury denied this up to his execution.) Five of the enslaved men were Spanish, captured by a New York privateer. The five men had been sold in New York as part of the prize, but they denied that they had been slaves while under Spanish jurisdiction and hence argued that they were being held unlawfully as slaves in New York. The accusation against the alleged priest and the Spanish captives coincided with a letter written by Governor Oglethorpe of Georgia and circulated to all colonial governors. In it he warned of a Spanish plot to infiltrate its agents into the British colonies, agents who would lay waste to colonial arsenals and principal towns, thus denying support for the British navy. The New York conspiracy was, Horsmanden believed, part of a larger military subterfuge in this current war. And, as the prosecuting attorneys argued to the juries, beyond that lay the two-centuries-old struggle between Protestants and Catholics. Thus was this uprising in New York mapped onto the wider history of religious struggle. It had been hatched by "the most abandoned whites, the dregs and disgrace of their complexion," with popery at the root of it all, declared Horsmanden. For Horsmanden and his fellow judges, lawyers, and jurymen, New York had become a stage upon which a historic struggle was being played out.[17]

If this interpretation had little role for slaves beyond carrying out the plots of others, nonetheless white New Yorkers expressed overwhelming fear and loathing toward the convicted slaves. The crowd at the execution of the first two convicted slaves was "very impatient to have done, their resentment being raised to the utmost pitch against them." Horsmanden struggled to explain how slaves whom he claimed were "treated here with tenderness and humanity" could partake in "so vile, so wicked, so monstrous, so execrable and hellish a scheme, as to murder and destroy [their]own masters and benefactors." These were no strangers, but people who lived in local households, were out and about town on errands, were familiar. The problem, Horsmanden believed, was not that slaves in New York suffered in their slavery, but indeed rather that they enjoyed an "excess of liberty." This explanation flattered slave-holding New Yorkers and absolved them of any responsibility for the conspiracy. "Their slavery among us is generally softened with great indulgence," claimed one of the prosecuting attorneys, William Smith, in his charge to the jury. "[T]he monstrous ingratitude of this black tribe" could only be explained by their own nature, that of a "brutish and bloody species of mankind."[18]

In this 1741 conspiracy all of the executed whites, even the accused priest who was believed to have been a prime mover in the plot, were hung. But almost half of the convicted slaves who were executed were burned alive. Burning at the stake, with its echoes of the exorcism of witches and heretics, was a gruesome death. It was a spectacular punishment meted out to those who threatened the social fabric more profoundly than did mere ordinary criminals. As such, it turned attention from any charge that it was white New Yorkers' own practices that threatened the social fabric, their binding of others to slave labor that was incompatible with social peace. An anonymous letter from a New Englander accused New Yorkers of hysteria for imagining such a vast conspiracy that 111 people were convicted, and likened it to the witchcraft trials in Salem in 1692. Indeed, while there is little doubt that some of the convicted indeed plotted something in New York in 1741, the fear the incident engendered and the venom with which vengeance was pursued reflected an animosity towards enslaved New Yorkers that was the culmination of a cluster of understandings of difference that had been developing in the colony for over a century.[19]

Efforts to deny a common humanity with slaves coexisted with the uneasy recognition of how thin the boundary between black and white actually was. Peggy Kerry, the woman who was among the first whites accused and executed for the 1741 conspiracy, had had a child with Caesar, one of the first slaves executed. In the rowdy public culture of New York, blacks and whites mingled at horse races, cock fights, impromptu dances, and fairs. On a couple of occasions, black women were reported to have given birth to twins of different colors, and mulattoes were constant evidence of how easily people mingled. After John Hughson was executed, his body was hung on a gibbet as a sign of his disgrace, and Caesar's body soon joined his on public display. A few weeks later rumor spread that "Hughson was turned Negro, and Vaarik's Caesar a white." Crowds flocked to the wall where they hung and indeed observed that Hughson's body was blackening, his hair was "curling like the wool of a negro's beard," and his nose widening, while Caesar's face was "somewhat bleached or turned whitish." Like the radish-hand dug up from a local garden and the Claverack bones, the cause of these transformations was the subject of much speculation, which no one could definitively resolve. Widespread witnessing of the transformation of the corpses stood in uneasy counterpoint to a racial idiom that claimed blacks were a separate "brutish and bloody species."[20]

*T*he War of Jenkins's Ear within which New Yorkers like Horsmanden set their explanation for their local conspiracy drew colonists out into imaginative engagement with the wider world in tandem with their actual engagement with this outbreak of another bout of imperial warfare. After 1739, when Britain and Spain commenced their low-level hostilities, rumors of a Spanish attack by sea periodically swept through New York, and New York ships commissioned as privateers sallied forth to capture enemy ships. While New York ships had privateered during previous wars, many more did so now. Then in 1744 Britain and France declared war on each other again, a war called in the colonies King George's War. Whereas during the two previous Anglo-French wars, New York had had no newspaper, by 1740 it had two and by 1743 three. These were now full of war news, allowing New Yorkers to follow Britain's efforts to defeat the French. And war news entailed geographic descriptions, in this case emphasizing descriptions of Canada. New Yorkers could also purchase a newly engraved Boston imprint, "A Plan of the City and Fortress of Louisbourg," the site of New Englanders' heroism during the war. Even more emphatically than had Britain's conflict with Spain, renewed war with France offered opportunities to visualize this struggle geographically.[21]

Still, even if Horsmanden and his fellow judges and lawyers interpreted events within New York as part of struggles between empires acted out on the world's stage, it is not clear that other colonists did. As with the two previous Anglo-French wars, many New Yorkers were generally disinclined to exert themselves on behalf of British imperial ambitions, even if they followed this one much more closely. Albany merchants still traded with the French in Montreal, as did New York City merchants, illicitly, with the French West Indies. Farmers feared being called away from their harvests for military duty; and everyone loathed taxes.

Despite local indifference, when the French and their Indian allies destroyed New York's northernmost settlement at Saratoga in 1745, pressure from London forced Governor George Clinton to marshal a response. He managed to assemble a force of militia from several colonies in the summer of 1746 for a march on Montreal. However, promised British regulars never arrived, New England's militias were called back, New York's Mohawk allies (already reluctant participants) complained that they hadn't been given promised rewards, and New York's own militia came to near mutiny over lack of pay. Clinton's force evaporated before it went anywhere, stymied by the Assembly's foot-dragging on raising revenue to support him, and his abandonment by others. Not much else happened in the colony for the rest of the war, with New Yorkers, as was so often the case, absorbed in local political squabbles.[22]

In King George's War, as in all colonial wars, Indians fought as allies on both sides. While New York newspapers occasionally reported without comment that New York's allies, the Mohawks, took scalps as did Indians allied with the French, one account of "French" Indians claimed that they mangled bodies and drank their blood. "This country," wrote Colden to one of his European correspondents "is now engaged in a most barbarous war with

Indians, popish converts, set on by accursed priests to murder innocent people." If many New Yorkers identified little with British imperial ambitions, anti-Catholicism united them across denominations and could thus be used to explain the savagery of particular Indians without tainting them all.[23]

But Horsmanden's framing of the 1741 conspiracy within a historic Protestant-Catholic struggle turned out to be tenuous and short lived. Africans had no visible association with Catholicism, and with the shift in enemy from Spain to France, middling New Yorkers revised their memory of the 1741 conspiracy so that it fit not within a narrative of Catholic-Protestant struggle but one of black rebellion. In the "Chronology of Things Remarkable" in his 1746 almanac, for example, Thomas More included "New-York's black Conspiracy detected." This understanding was continually affirmed by repeated newspaper reports of such conspiracies elsewhere (in contrast to an incident in which Irish servants allegedly conspired to rebel in Philadelphia, which was reported as if they were simply silly). Ignoring the free whites and the enslaved Spanish and Indians who were among the convicted, the conspirators were now simply "black," with the 1741 conspiracy framed within a narrative of a chronic black threat.[24]

𝒜s the tide of war news receded, geographical information and maps were more visible in New York than they had been before the war. Throughout the British Atlantic over the eighteenth century, the mastery of geography became more closely associated with refinement and improvement. So too, slowly, in New York. By 1749 the shopkeeper Gerardus Duyckinck was advertising "seascapes, landscapes, prospects . . . and maps of the world," and James Parker had added geographies to the books he offered for sale. The winter of 1748–1749 was when Mr. Bonin showed off his optical machine with its "Perspective Views of the most famous Palaces and Gardens in England, France, and Holland." Samuel Johnson, the Anglican minister who would become the first president of King's College, ran a school for young gentlemen in the early 1750s that offered "modern geography with an account of the present state of the kingdoms and republics in Europe and the great monarchies in other parts of the world." Newspapers kept readers abreast of news of the boundary negotiations in Canada, the search for the Northwest Passage, and discoveries in the South Seas. Such news promoted a panoptic grasp of the world that characterized refinement and served as evidence that the British were in the forefront of improvement.[25]

Reciprocally, as the rest of the world became more visible, increasingly so did New York. In 1750 the printer Henry De Foreest began offering

"The New-York Primmer." In 1753 three young men from prominent New York families, William Livingston, William Smith Jr., and John Morin Scott, began publishing a weekly newsletter, *The Independent Reflector*, with the intention of bring to New York a "polite" publication modeled after London's *Tatler* and *Spectator*. Such a publication would allow for a critical eye to be turned on affairs in New York itself, and, the young men hoped, make New Yorkers more conscious of the need for improving their colony. While it lasted little more than a year, among other essays the young men offered reflections on their native place. "A Brief Consideration of New-York" lauded New York's soil, harbors and rivers, opportunities to produce naval stores, potential for mineral strikes, and its seafood (especially its oysters). While the trio found much that needed improving, their enunciation of the particularities of New York and their comparisons of it with nearby colonies could make readers more conscious of the distinctive characteristics of New York. In 1751 Thomas More offered his almanac readers a chronology of New York's governors, beginning with the English seizure of New Netherland in 1664. If the panorama of the world's places was becoming more visible to New Yorkers by the 1750s, so too were they becoming more visible to themselves.[26]

One cannot make too much of this, however. If some New Yorkers beyond the handful of officials with political interests in geography began to imagine New York as a singular entity, most people were more likely to continue to identify themselves with their local communities and with their ethnic and religious affiliations. When three men living in the Mohawk Valley wrote Cadwallader Colden in 1753 of their need for a surveyor, for example, they referred to "your Yorck Gentlemen," as if they didn't think of themselves as particularly living in New York. Dutch New Yorkers were conscious that English New Yorkers sometimes mocked them, and English New Yorkers complained that minor officials who were Dutch showed partiality towards their own. Rural Dutch, in particular, lived as if British claims to sovereignty were of little moment and, in fact, the Classis of Amsterdam, the ecclesiastical body that supplied the Dutch Reformed Church with ordained ministers, continued to refer to the area as "New Netherland" into the 1750s. James Parker printed an account of heroism at Louisbourg in 1747, and the Zengers printed "The cries of the oppressed, or Herods cruelty displayed; at the taking of Bergen op Zoom by the French" for their Dutch customers in 1748. An account of the 1747 storming of this Netherlands city as part of the war in Europe, the pamphlet evoked Dutch narratives of their valiant Protestant resistance to their Spanish Catholic overlords during their long war for independence (1579–1640), with the French now replacing the

Spanish as the Catholic Antichrist. Thus, different reading publics were offered interpretations of the significance of King George's War, readings that
affirmed different identities. Still other New Yorkers—German, Huguenot,
Africans, Indians—would have had little reason to identify with Britain's
struggle nor have had an interest in reimagining New York.[27]

Nonetheless, in this context of the gradually emerging visibility of
new practices of geography and cartography, maps that showed New York
itself began to be offered for sale more frequently. More and more anchored
in actual measurements, these maps offered the kind of detail that increasingly signified legitimacy. However, even though there was clearly a market
for these maps in the colony, New Yorkers did not mistake the accuracy with
which natural features were represented with disinterestedness.

The first of these was a map printed in 1749 by Pennsylvanian Lewis
Evans, a "Map of Pensilvania, New-Jersey, New-York, And the Three Delaware Counties." To make his map Evans had collected information from
anyone likely to have measured the latitude or longitude of points in the region, or the distance and direction between any two points. He had consulted
maps printed in England, and corrected and augmented their information
with observations of his own. He himself had been hired to survey parts of
Pennsylvania, had taken observations on a trip to Onondaga and more on
a trip to Albany, and had assisted James Alexander, the surveyor general
of New Jersey, in making maps of an area there whose title was in dispute.
Thus he was able, more than anyone else before him, to make a map based
on actual measurements of the region.[28]

Still, some were quick to dispute the political assertions on his map.
"A.B." from Orange County on the Hudson just north of New Jersey objected to Evans's map on behalf of "myself and other freeholders." The New
Yorker claimed that Evans had laid down the New York–New Jersey boundary line incorrectly, a boundary that had indeed been left up in the air after
the abortive 1719 attempt to agree on it. Evans's critic objected to the way
Evans's map served some people's interests by drawing a politically created
boundary line with the same assurance as he had drawn natural phenomena
like rivers, thus making the boundary line seem as natural. Evans defended
himself, describing the sources he had used and asking his critic to share any
evidence he had that the map was in error. He also asserted that he had "no
Interest in this Controversy," and by inviting his critic to provide documenta-
tion for the freeholders' version, Evans was proceeding as if boundary lines
were objective phenomena whose true nature could be determined through
the disinterested weighing of evidence. And indeed in 1753 Evans published

a new edition of his map, with new information and corrections of past errors (although the boundary line stood, his critic apparently unable to supply him with evidence that he had erred). Evans made no distinction between the accuracy of natural and political phenomena when he announced that his map had "a Degree of Correctness that the Maps of but few of the most civilized Parts of Europe have arrived at."[29]

In 1755 Evans published another map, *A general Map of the Middle British Colonies in America* (illustration 9). On it he showed all the information he had amassed about Pennsylvania, New Jersey, and New York, this time adding Maryland and much of Virginia. But the settled areas of the colonies occupied only the right half of the map. On the left was territory to the west, including the Great Lakes and the Ohio Valley. Thus this map made visible to readers the relationships of their colonies to the interior. Just as New York officials had long claimed to their London superiors, Evans asserted that all this territory was under the sovereignty of the Iroquois, printing "Aquanishuanigy" in a sweep from the left edge of the map to just south of Lake Champlain in the upper right. That many of the literate in New York indeed had only the haziest of notions about this geography is indicated by a comment by Goldsbrow Banyar, a second-level official who had lived in the colony since the 1730s. "I never knew before I saw Evans's Map where the Beaver Hunting was [the subject of the Iroquois' 1701 deed]," he marveled, "and little expected to find it there, on the North side [of] Lake Erie."[30]

But despite Evans's repetition of the claim of the Iroquois to a vast expanse of the interior, his map did not otherwise support the arguments of New York officials. Instead of affirming the importance of the Iroquois' claims to part of Canada and the consequent significance of New York, on both the map and in an accompanying pamphlet Evans stressed the desirability of the Ohio Valley and the proximity of both Pennsylvania and Virginia to it. Evans's map made it seem easy for Pennsylvanians to get to the Ohio Valley via the Susquehanna River and Virginians to get there via the Potomac. Only invitingly small spaces separated the heads of these rivers from branches of the Ohio. The map, then, supported other colonies' arguments for their importance to the empire, rather than that advanced by New York.[31]

As it happened, Evans published this map as Britain and France were preparing yet again to go to war. Initially the ministry in London directed the British military commander, General Braddock, to indeed concentrate on the forks of the Ohio. Relying at least partially on Evans's map, Braddock followed the Virginia route to the Monongahela and met disaster. With Braddock's failure and death, Governor Shirley of Massachusetts inherited

ILLUSTRATION 9. Lewis Evans, *A general MAP of the MIDDLE BRITISH COLONIES in America* (1755). This map made broad claims for Iroquois territorial sovereignty as encompassing the entire Ohio Valley, as well as territory north of Lake Erie. *(Collections of the New-York Historical Society.)*

command and changed tactics. As New York officials had long argued, Shirley believed that military resources should be concentrated in the north. Consequently he proposed a multi-pronged strategy, with attacks on French forts at Crown Point north of Albany, Frontenac on Lake Ontario, and Niagara. At last, it seemed, New York would be the focus of imperial attention and acknowledged as key to Britain's empire in North America.[32]

But over the winter of 1755–1756, warfare endemic to politics in New York erupted over Shirley's strategy. Crown Point and Niagara were on the edges of Iroquoia, the territory between the Adirondacks and Lake Erie that the Iroquois controlled, and thus plausibly under British sovereignty via their Iroquois alliance. Frontenac was across Lake Ontario. On his map Evans had indeed shown the first two as within territory that could be claimed as British, but he had also shown Frontenac as within the bounds of French Canada. New York opponents of Shirley now held up Evans's map as evidence that Shirley's plan was illegitimate. In so doing, of course, they seemed to be undercutting a long-standing argument for New York as the rightful focus of the struggle against the French. But the advocacy of New York's special geography had been the work solely of a handful of royal officials. As we have seen, few New Yorkers had shown much enthusiasm for these anti-French wars. And in any case, Shirley's opponents had only a passing allegiance to the pronouncements on Evans's map. Rather, their interests were at stake: Shirley had granted contracts for military supplies to their local opponents and he had banned the Albany-Montreal trade for the duration of the war. For Shirley's opponents, Evans's map was merely a handy weapon.[33]

Supporters of Shirley then counterattacked by decrying Evans and his map. There should be a law, one such supporter proclaimed in a letter to a local newspaper, "to restrain these Gentlemen, who divert themselves with setting Bounds to Provinces and Empires, 'till their Works have stood the Test of an accurate Examination." The supporter, Roger Mortar of New York City, warned readers that as instruments of persuasion, maps could be dangerous. The danger of Evans's map, he wrote, was that it might make soldiers reluctant to fight, discourage support for the Empire within Britain, and hearten the French. As evidence for maps' persuasive powers, Mortar pointed out that "it is well known, that not long since, a very respectable Body, unacquainted with his Majesty's Rights in America, or influenced by false Geographical Representations, seemed in an Answer to a Message from one of our Governors, to doubt the Equity of our Claim to the very Lands on the Ohio." (In 1754 a member of Virginia's House of Burgesses had indeed used DeLisle's map as justification for declining to support Governor Robert

Dinwiddie's request for funds to support a military excursion to the Ohio.) Maps like those of Evans were dangerous, Mortar argued, because of the risk that map readers would failed to recognize the interests served by the map.[34]

But here Mortar was guilty of some hyperbole. Whatever claims maps made to disinterestedness through the measured location of physical landmarks and their careful representation on paper, neither the Virginia representative nor Shirley's New York opponents mistook the reliable representation of natural features on maps for the absence of any political assertions. And yet the very debate showed that this understanding of maps as panoptic, disinterested, and reliable representations of the world "as it really is" was gradually rising in people's consciousnesses, even if Mortar considerably exaggerated the extent to which people were gullible enough to believe maps were simple representations of fact.

If King George's War raised the visibility of geography in New York, the Seven Years War raised it further, and dramatically. Iroquoia and the northern borderlands of New York were the sites of much more military action than had ever been the case previously. In September 1755 William Johnson led a force of militia and Iroquois into victory at the Battle of Lake George, a battle immediately celebrated as heroic. The next year the tides of war turned, with the French capturing the fort at Oswego on Lake Ontario. In 1757 they seized Fort William Henry, and in the aftermath surrendering British regulars, colonial militia, and Indians were all massacred. A British assault on the French fort of Carillon on Lake Champlain failed in July 1758 with horrific losses, but in August the British took back Oswego and then took Fort Frontenac across the lake. Finally, in 1759 the British defeated the French around Lake George and Lake Champlain, while in western Iroquoia a force led by Johnson took the French fort at Niagara.[35]

Although New Yorkers' support for previous wars had been tepid, this time the Assembly more than tripled the funds they had appropriated for King George's War, and several dozen New York ships sailed out as privateers. Not only were all these events of war extensively reported in New York newspapers, but they were also described in local almanacs. John Nathan Hutchins's account of the loss of Fort William Henry in his 1759 almanac ran to six pages. And these accounts were accompanied by geographic descriptions of the relevant territory, with the *New-York Weekly Post-Boy* advising readers to acquaint themselves with geography so they could follow the news of war.[36]

New Yorkers could in fact more easily acquaint themselves with geography because maps proliferated. By now there were a half-dozen mathematics schools in and around New York City, so more young men were trained in surveying. The British army now had a corps of engineers that dispatched men to survey routes for marching troops, lay out trenches for sieges, and chart waterways. A half dozen of these engineers circulated through New York during the Seven Years War, some staying on afterward. In short, as the demand for maps rose, so did a cadre of men capable of responding to that demand. In 1756 one Samuel Blodget offered his engraving for sale, a "Prospective Plan of the Battle near Lake George," so New Yorkers could see where Johnson had enacted his heroism, a map well advertised in New York. In 1758 both the *New-York Mercury* and the *New-York Weekly Post-Boy* printed a map of the fortress of Louisbourg that had just been retaken by the British. In 1760 Thomas More included a woodcut map of Quebec with his account of the course of the war for his almanac readers. Evans's map was reissued in 1760 to show sites newly imbued with significance by the events of the war. New York newspapers reported admiringly of British mapping work in the region, from which, one wrote, New Yorkers could "obtain a certain knowledge of our own Frontiers, which, 'till very lately, we knew little of, but from the French Accounts."[37]

"All Canada In the Hands of The English" trumpeted the headline of the *New-York Weekly Post-Boy* in September 1760. As was true throughout the British colonies, the vicissitudes of this war dramatically increased expressions of loyalty to and identification with Britain, with a corresponding demonization of the French. In his 1759 almanac, John Nathan Hutchins described the French as "a restless and never-to-be-satisfied Enemy, who has ever since the Year 1689 . . . laid Siege to the Province of New York." Thus Hutchins wrote the history of the colony as one in which it had been continually struggling against the French on its border, even though that had not been altogether apparent to most New Yorkers. Roger More headed all the calendar pages of his 1761 almanac with verses on the glorious victories of the British over the French, and Thomas More congratulated his almanac readers on "the Triumphs of our Nation."[38]

Even more markedly than in the aftermath of King George's War, after the Seven Years War geographical descriptions and cartographic representations gained a visibility in New York far beyond the maritime maps long purchased by sea captains and merchants. By the 1760s booksellers consistently offered an array of geographies and atlases, and dozens of maps of places near and far. And just as New Yorkers could now venture imaginatively to distant places, so too increasingly could they contemplate New York. New geographies like the *American Gazetteer* described over two dozen New York towns, counties, and natural features. William Smith Jr., one of the trio who had published *The Independent Reflector*, published a history of New York in 1758, a history that was accompanied by an assortment of maps. Booksellers and shopkeepers in New York City now advertised an array of engraved prints of local landscapes like the falls on the Mohawk River at Cohoes, the Tappan See, a widening of the Hudson framed by magnificent cliffs, and New York City itself as seen from the harbor. Newspapers also carried more news of other colonies, sometimes with explicit comparisons with New York, and booksellers offered titles like Peter Kalm's *Travels in North America*, wherein New Yorkers could read observations of themselves alongside observations of other colonies. If the war had excited some New Yorkers to engage imaginatively with a far-flung British Empire, it simultaneously heightened still further the visibility of New York itself.[39]

This visibility of New York included maps that were based on the work of the cadre of surveyors, both military and lay, that since the onset of the Seven Years War measured more and more of the topography of New York. A map printed in 1776, Claude-Joseph Sauthier's *A Map of the Province of New-York, Reduc'd from the large drawing of that Province, compiled from Actual Surveys*, exemplifies this new sort of map (illustration 10). Here New York was just itself, completely enclosed, a place with clear, fixed boundaries. Unlike maps like the ones sent by Bellomont and Burnet in which New York was significant because of its relationship to somewhere else, other places simply surrounded the colony, anchoring it in place but having no obvious significance in themselves. And unlike maps even as recent as Evans's, New York was not a general area but a clearly bounded space. Indeed, boundary commissions had finally agreed on the New York–New Jersey boundary in 1769, and on the boundary line with Massachusetts in 1773. A treaty with the Iroquois in 1768 had fixed New York's western boundary, and the Crown had recognized New York's jurisdiction north of Massachusetts as far as the Connecticut River in 1764, and had clarified the colony's northern border with the Quebec Act of 1774. The clarity and accuracy with which natural features were represented on this map were echoed by the clarity with which its political features were laid down.[40]

*H*owever serene and orderly New York might have appeared on Sauthier's map, the colony was anything but calm in the thirteen years between the end of the Seven Years War and the beginning of the Revolution, and land was a major source of contention. Like a cork out of a bottle, the removal of

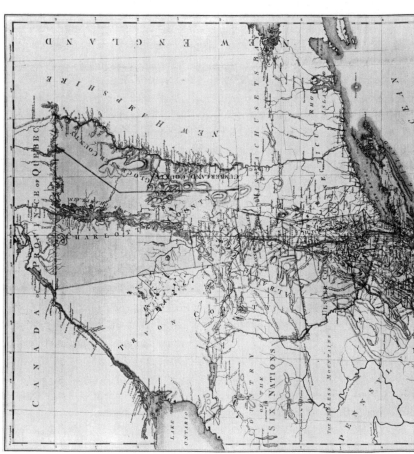

ILLUSTRATION 10. Claude-Joseph Sauthier, *A Map of the Province of New-York, Reduc'd from the large drawing of that Province, compiled from Actual Surveys* (1776). Bottom quarter, with cartouche and southern New Jersey omitted. In Sautier's depiction, New York is a colony with clear boundaries that is internally organized into counties of similar clarity. All of Cumberland and Glocester Counties and much of Charlotte County was claimed by settlers from New Hampshire, however, and Tryon County encroached on Mohawk territorial claims. (*Collections of the New-York Historical Society.*)

the French as a military threat from New York's northern borders released pressure to expand settlement. Just before the Seven Years War a census counted 97,000 New Yorkers, black and white; fifteen years later in 1771, a census counted 168,000 residents. Hundreds of newly discharged British soldiers took up warrants for land north of Albany, as did several thousand new British colonists. Moreover, wealthy Britons who stayed home but were looking for investment opportunities now turned their eyes to New York. As Pennsylvania's Indian agent George Croghan wrote William Johnson, "one half of England is Now Land Mad & Every Body there has thire Eys fixt on this Cuntry." As importantly, New Yorkers began to lay claim to land titles that they had been accumulating since the late seventeenth century but often enough had left un-surveyed and with their exact boundaries obscure. In the land rush that erupted in the wake of the war, New Yorkers wanted both to secure final title to land within the colony itself and to land supposedly long off limits as falling within the territory of the Iroquois.[41]

The Seven Years War brought a marked change in white New Yorkers' representation of Indians because it was now in their interests to do so. Roger More wrote a verse about Fort William Henry for inclusion in his 1758 almanac that read in part:

> Those hell-hound Savages, with cruel Joy,
> Glutted with Blood, the Helpless there destroy.

While Indians who were enemies in previous wars had certainly been decried for their barbarism, such characterizations had been only episodic. With the Seven Years War, such characterizations were more sustained and seemed to encompass Indians as a singular and unified group. Indians in general were described in a magazine article as characterized by "inhuman and barbarous Customs they have always been subject to, namely human Sacrifice and putting their Prisoners to Death with the most exquisite Tortures and unprecedented Cruelties." With the war's end these characterizations of Indians as savages persisted, and, importantly, now included the Iroquois, because with the French gone so too was the value of the Iroquois. Oneidas, for example, were reported to be preparing to roast some enemy Indian captives "agreeable to Indian custom." In this newspaper article, the Oneidas were referred to as "feroce Bipeds," in short as animals. They were wild as animals were wild, one with an untamed wilderness. More erudite New Yorkers cloaked their self-interest in Indian savagery by drawing on nascent Enlightenment representations of the noble savage. They are, the physician Peter Middleton blithely asserted in his address at the opening of

King's College medical school in 1769, utterly untouched by civilization. Indians, wrote William Smith in his history of New York, were completely without agriculture, utterly free of restraint (although he conceded that Indian women "cultivate a little spot of ground for corn"). More modest New Yorkers were no less prone to disguising their self-interest by asserting that Indians were wilderness savages. Like Copernicus Weather-Guesser's verse that opened this chapter, they rewrote themselves into a story in which they erected civilization by subduing wilderness. Whether Indians were savages like wolves, thus needing to be exterminated, as in Weather-Guesser's formulation, or part of the savage landscape of a more romanticized nature that would naturally yield to civilization, New Yorkers' embrace of the metaphor of savagery allowed them to disguise their own ruthlessness in dispossessing Indians from their land.[42]

In the wake of the war, the king's ministers in London greatly increased their attention to their North American colonies, and particularly to the problem of cultivating peaceful relations with Indians. Faced with Indian hostility to settler encroachment, mounting debts for military actions in North America, and the rising clamor of Indian complaints of land fraud, London authorities began an effort to soothe Indians by protecting their land. In 1761 the king ordered that all land grants in New York beside the Mohawk River and up around Lake George cease until Indian complaints were satisfied. After peace was concluded with the French, the Crown decisively ordered New York's Assembly to void the Kayaderosseras Patent, an early eighteenth-century land grant revealed as containing hundreds of thousands of acres that stretched north and west from the confluence of the Hudson and Mohawk Rivers, but that only now seemed safe enough to settle. The Mohawks, who had been among the most reliable of British allies, complained vociferously about it, asserting that they had never sold such a vast expanse of land. The Assembly, however, refused to annul it. The Assembly claimed that Indians often deny the sales of their ancestors, and anyway the Mohawks probably just wanted to be paid over again. In addition, they piously expressed concern that voiding patents would "arraign the conduct of former Governors" (although they had never previously been particularly solicitous of the honor of their governors). "[I]t is impossible," the Assembly asserted, "to discover the true Owners of any lands among unlettered Barbarians, who keep no certain Memorials; have very indistinct Notions of private Property; live by hunting; use no Land Marks; nor have any Inclosures." The Kayaderosseras Patent, originally claimed by a handful of people, had by now passed to well over a hundred of their descendents, many of whom were members of the most prominent families in New York. If they were to have their land, Indians needed to be savages whose words were meaningless.[43]

In 1764 yet another claim of fraud came to the attention of the Crown. The Wappingers were Indians who lived in the mid-Hudson Valley. When the Seven Years War began, the Wappingers left their land, with able-bodied men going off to fight the French and the rest removing to the Indian town of Stockbridge in the Berkshires for the duration of the war. When they returned at war's end, they were informed by the heirs of Adolph Phillipse, one of the recipients of the large grants conferred by Governor Fletcher, that the roughly 25,000 acres the Wappingers considered their own was in fact included in Phillipse's 1697 deed. No Wappinger was alive who had witnessed the 1697 transaction, and the dimensions of the land that were written on the paper deed had been kept secret, no doubt because, as William Johnson speculated, "the original Patentee not daring to let them know, how much was contained in his grant, until they dwindled to an inconsiderable Number." The Wappingers sued in court and lost. No New Yorker was going to threaten another's title by finding against him in court, for his own title might be just as weak. The Wappingers then petitioned Colden as acting governor and, unsurprisingly, lost again. "The Indian Nations are a mere mob," protested Colden to his superiors in London who were exerting pressure for justice for the Wappingers. They are "directed by popular leaders who are governed by their passions with a violent inclination to war, and easily instigated to revenge, which is the characterestick passion of Savages." (This from a man who in the 1720s had sought British acknowledgement of New York's importance by describing the Iroquois as "noble as Romans.")[44]

In despair, seven Wappingers traveled to London to seek the Crown's assistance directly. The Lords of Trade concluded that their complaint warranted further investigation and directed the incoming governor, Henry Moore, to do so. "I shall always think it my duty to protect them in all their Rights and Privileges and to discourage every appearance of a Design to oppress them," wrote Moore after his investigation. But he could get nowhere in New York. The Governor's Council, like the court, was implacable in its support for the Phillipse claim. And so, "for the sake of Peace and good Order," Moore let the Phillipse patent stand. The Wappingers were dispossessed. As the Assembly wrote, "the Precedent of annulling Crown Grants . . . will render all Property insecure; give highest Dissatisfaction to His Majesty's subjects, alarm their Fears, and discourage the Settlement, Cultivation, and Prosperity of the Colony." Moore's successor, Governor William Tryon, would go further, denying that Indians ever had had title to land in or near the colony; paying Indians for land was merely to buy their quiescence. No one's interests were served any longer by acknowledging Indian rights.[45]

But if land-owning New Yorkers, and those with aspirations for land, were now united in their characterization of Indian savagery, they were not necessarily united on other fronts. However much Sauthier's map had the potential to mark off New York as a distinct space, New Yorkers knew full well the politics of the map. Sauthier, while trained as a landscape architect and a surveyor and thus having the skills necessary for mapmaking, had come to New York 1771 as an aide to the new governor, William Tryon. On his map, Sauthier had included the region of the Green Mountains east of Lake Champlain as part of New York. But both civil jurisdiction and property rights were strongly disputed in the area. New York's claims to it were contested by the white occupants, most of whom held grants given by New Hampshire, which claimed its jurisdiction ran as far west as did that of Massachusetts. Claiming New York's rights to the area, Tryon had granted himself more than 30,000 acres in the region, as well as granting 5,000 acres to Sauthier and thousands more to others. In the face of violent opposition from the New Hampshire grantees, the New York titles were just paper. Members of the Assembly had refused to raise money for a militia to enforce titles in which they had no share. Crown officials similarly had refused the use of British troops, for although the Crown had confirmed New York's claims to the Green Mountains in 1764, the Board of Trade had ordered New York to cease granting land in the area in 1767 until disputes over title could be resolved. Tryon had also acquired, with William Johnson's help, more than a million acres for clients (and 40,000 acres for himself) from the Iroquois. Johnson (whose own interests in acquiring land were poorly understood in London) had negotiated the 1768 agreement with the Iroquois that had allowed Sauthier to draw such a clear western border for the colony. But in so doing, Johnson had incorporated land previously recognized as part of Iroquoia into New York, and thus available for purchase, incurring the displeasure of the London ministry.[46]

In all of this Tryon made no pretence of holding land grants—to himself or to favorites—within the 2,000 acre limit that had ostensibly been imposed on New York in 1709. To rebukes from London about such grants, Tryon tried to claim that granting land to "gentlemen" would "create subordination and counterpoise, in some measure, the general leveling spirit" in the colony, an argument hardly popular among more ordinary New Yorkers. Nor was Tryon's argument well received in London. At the end of 1772

Tryon was ordered to stop granting land in New York altogether. The map Sauthier produced, then, was one that tried to naturalize in cartographic representation the boundaries of a colony that were far from accepted on the actual ground. In New York the representation of land in maps was never mistaken for the land itself. Maps were not understood as simply objective and uninflected reflections of territory, for the stakes in their representations were too obvious.[47]

*B*oth Indians and slaves were by now consistently represented as savages, despite the evidence of white New Yorkers' own experiences. From Montauk to the Mohawk Valley, Indians had long incorporated aspects of colonists' agriculture into their own practices, engaged in trade, and worked for wages. And slaves led lives in New York much like everyone else. Because of the centrality of commerce to the colony, a number of slaves were taught to read, write, and do the arithmetic usual to commercial transactions because their owners needed such skills from them. Enslaved men mastered all the common trades of the colony: carpentry, masonry, shipbuilding, rope making, cooperage, and smithing. Although new slaves were brought continuously into the colony from Africa as well as the Caribbean for the entire colonial period, the majority of slaves by the mid-eighteenth century had lived for a good while in the colony, with some born there and a few descending from people who had been there for generations. To some variable degree, those with African and Indian origins remained ethnically distinct—that is, with characteristic practices in food, dress, music, and recreation. But this was not so very different from New Yorkers with Dutch, French, English, and German origins who continued ethnically distinctive practices in the eighteenth century. By the 1760s slaves held by Dutch New Yorkers in the Hudson Valley, who had long joined their owners in the festive celebration of Pentecost (called in New York "Pinkster Day" after a local honeysuckle that blossomed then), had added their own component of African music and the crowning of an African "king." Nonetheless, as Albany shopkeepers' records and ads describing runaway slaves show, slaves throughout the colony engaged in the same sorts of activities of ordinary daily living as did whites.[48]

One of the deep ironies of the slow but clear shift from conceptualizing Indians and Africans as heathens to insisting that they were savages is that the Indians and Africans in and near New York were now increasingly likely to be Christians. Seventeenth-century New Yorkers had been complacent about their rights to exclude Indians from political and legal rights and to hold Africans as slaves because as non-Christians they were outsiders, but that argument did not hold any longer. Conversions of handfuls of Indians and Africans had begun in the seventeenth century, and with the Great Awakening of the 1740s more and more of them found spiritual sustenance in Christian belief. Jupiter Hammon, a slave held by the Lloyd family on Long Island, wrote devotional poetry that reflected his deep religious sentiments. David Fowler, a Montuakett, welcomed the ministry of Samson Occom, himself a Mohegan trained by Eleazar Wheelock, when Occom evangelized on Long Island. Slaves became members of the Anglican Church in New York City, and John Ogilvie, an Anglican minister in the Mohawk Valley in the 1750s, recorded the baptism of babies white, Indian, and black. Mary Cooper, the Long Island farm woman, noted in her diary an "Indian preacher in town," and another time commented, "I went to the New Light meeten to here a Black man preach." In these white accounts, blacks and Indians were always marked as such, in contrast to whites whose activity was recorded without remarking on their color, so the conviction of their outsider-ness was deeply held by whites. Nonetheless, many of them were no longer even remotely heathen, but spiritually like white New Yorkers. If white New Yorkers deceived themselves about the nature of Indians and Africans, marking them as unalterably outside the social body and the body politic, they did so because it was so powerfully in their own interests to do so. Only with savage Indians and savage slaves could New Yorkers be justified in taking the land of one and the labor of the other.[49]

This two-pronged movement, simultaneously embracing a taxonomy of natural human difference while remaining wary of the seemingly objective presentation of place on maps, again points to how critical social relations were to knowledge and to the degree to which people were willing to change their minds about what they knew. Colonists asserted that they knew the natural world through experience, and they saw no contradiction between what they believed they knew through physical interaction with the material world and how they interpreted those experiences using familiar texts like the Bible and classical sources. When new assertions about what the natural world was really like were presented to New Yorkers, their willingness to consider them was strongly colored by the degree to which they believed the new assertions served others' interests but not their own. Colonists were persistently anxious about other people's pursuit of self-interest in public matters, which political discourses show, particularly in the decade before the declaration of independence from Britain. In New York, by and large, it was the political elite who had a stake in having maps accepted as straightforward depictions of the colony. But political fractures within the colony were too strong and too evident for maps ever to become understood as simply representations of what was actually there, even if maps were considerably more common than earlier.

However divided literate New Yorkers may have been among themselves over a host of issues, their common interests in African labor and Indian land made it easy for them to exclude Africans and Indians from being New Yorkers. If the meaning of the physical landscape was still a matter for dispute, the meaning of the human one, white New Yorkers hoped, had been settled.

Conclusion

Nature and Knowledge in New York

On the eve of the Revolution New Yorkers' knowledge of the natural world was somewhat different from what they had believed and practiced in the seventeenth century. By the 1760s and 1770s, ordinary New Yorkers were more willing to consider new material innovations, for their own experiences with improvements like hemp cultivation, more precise navigational instruments, and smallpox inoculation led them to be more optimistic that they were living in a time of progress. They were more aware of what constituted a refined apprehension of the natural world, that an elevated character could be demonstrated by displaying elegant exotic plants in a garden or through coolly contemplating phenomena like lightning or comets that frightened the unenlightened. At least some grew confident that reason told them that magic and witchcraft did not exist, but were mere superstitions, and even those who continued to believe in them were now reticent about those beliefs in public. No doubt clusters of people in rural settlements or among the marginal in towns remained unaware of some or many of these changes, but newspaper and almanac readers could not but acknowledge them.

Nonetheless, this is not a story in which New Yorkers simply learned a truer knowledge of the natural world. As the story of Joseph Morgan shows, it was certainly possible by the early eighteenth century for even a modest rural minister to embrace inventiveness, learn of the geography of elsewhere, elaborate on Newton, promote public improvements, and at least attempt to correspond with the Royal Society. Yet interest in much of this was limited among most New Yorkers right up through the Revolution. Certainly some New Yorkers—Cadwallader Colden with his natural philosophy, Jane Colden with her botanical work, Samuel Skinner with his comets, Claude-Joseph Sauthier with his map—actively engaged in pursuits we now deem scientific. They are evidence that new ideas and practices circulated easily in the Atlantic world, but also that the simple presentation of new assertions about the natural world was not enough to convince everyone that they were true. The literate but not learned who were newspaper and almanac readers in New York were more skeptical of new truth claims, more wary of innovation, and more resistant to acknowledging the authority of others' knowledge than were our exemplars of new beliefs about and practices in the natural world like Morgan, the Coldens, Skinner, and Sauthier.

There are several reasons for this halting and uneven shift in knowledge. For one, people understood that much of their knowledge of the natural world was based upon experience, which shaped what innovations they would or would not accept. People who labored with soil, crops, and animals, with wood, metal, and cloth knew what they knew through their senses. This is not to say that they knew things in some primary, pre-verbal way. As the chapters in part 1 should have made clear, their experiences were understood through specific cultural frames. Their conceptions of the human bodies as a sac of fluids whose balance and circulation constituted health and whose skin was permeable in both directions was affirmed by the material experience of leaking, stinking, feverish, and painful bodies and of their bodies' clear responsiveness to cathartics, emetics, bleeding, and blistering. The variability of weather and climate mirrored the ever-changing patterns in the night sky. The inconstancy of the natural world and the uncertainty of the results of people's efforts meant that approximation and trial and error were familiar parameters within which they struggled.

When change came, the innovations clustered in part 2 under "improvement" were most amenable to personal trial and easiest to accept. Those clustered under "refinement" held an intermediate ground: available to sensation but less central to everyday existence. Those clustered under "reason" were least amenable to personal experience. Copernican cosmology and Newtonian celestial mechanics did more than merely redescribe how the universe worked. They also described a cosmos that could not be affirmed by ordinary people's experiences, a cosmos that they could not know for themselves but could only know through trusting the assertions of the erudite. And that many were unwilling to do.

As the last chapter shows, experience was itself a flexible claim. White New Yorkers increasingly ignored the evidence of their own experience when considering the Indians and Africans among them. In the seventeenth century, colonists and Algonquians, Hudson River Indians, and Iroquois learned, to some degree, the knowledge held by the others. Colonists learned the lay of the land, the location of shellfish beds, skills in whaling, and techniques for making rope and canoes. Indians adopted new crops, livestock, and material goods. These exchanges more and more took place within a framework of steadily increasing political and social inequality, a development that the Iroquois, in possession of their own territory and politically valuable, largely escaped until after the Seven Years War. When the defeat of the French robbed the Iroquois of their importance, abruptly white New Yorkers ascribed to all Indians a natural savagery that erased their rights to land. Henceforth, all Indians were represented as mere nomads, wandering over the earth just as animals did, against the evidence of experience. So too did white New Yorkers come to mark off Africans and their descendents in the colonies as natural savages. Initially in the early seventeenth century, the boundaries between African and European, and between slave and free were fairly permeable. But colonists by the late seventeenth century had embraced the hardening of slavery that occurred everywhere in the Atlantic world. The recurring rebellions and resistance of people enslaved in the colony were henceforth ascribed to, as the lawyer William Smith put it, their "brutish and bloody" nature. This too was against the evidence of experience, for slaves in the colony worked, ate, played, and prayed alongside whites. Insistence on the difference of Indians and Africans flew in the face of white New Yorkers' own experiences of what were in fact very modest differences.

A second reason for the slow acceptance of new assertions about the nature of the natural world was New Yorkers' widespread resistance to changes in the social relations of knowledge: who was acknowledged to possess knowledge, how it circulated, on what grounds it was accepted. The familiar economy of knowledge was one in which most people shared most sorts of knowledge about the natural world and acted upon that knowledge in their everyday lives. They did most of their own healing, dealt with birthing or ailing livestock, nurtured gardens, orchards, and fields, fought infestations of rats and locusts, coordinated travel, fishing, and shellfish gathering with the tides, and watched the phases of the moon. When they drew on the knowledge of others, such as a surveyor or a midwife, those others possessed knowledge that was merely an accentuation of what everyone knew. This is in marked contrast to modern regimes of knowledge, one in which a plethora of people with specialized educations—engineers, physicians, MBAs, college professors—dispense information and advice authoritatively to those uneducated in their field. In the eighteenth century New Yorkers well recognized that such claims to enlightened learning were simultaneously claims to elevated social authority. As with the physicians who tried to elevate their standing, Colden who tried to out-do Newton, and Frank Freeman who tried to persuade his readers to abandon astrology, all of these were claims that some people really knew the natural world and that others were simply ignorant. In a colony in which people recognized sharply how expressions of political or religious ideals were often merely window dressing for personal ambition, New Yorkers were generally unwilling to concede authority to those who expressed allegiance to erudite realms of knowledge of the natural world. Acutely aware of others' self-interest, New Yorkers quickly recognized that assertions that were presented as objectively true, like Sauthier's map, often promoted particular interests, like Governor Tryon's property interests. At the same time, people easily disguised from themselves their own self-interests, as the tale of the hardening conceptions of race among white New Yorkers shows. While some people certainly embraced aspects of the new learning because they were convinced that it was more useful and more accurate than older ways of understanding the natural world, nonetheless the simple demonstration of usefulness and accuracy was not enough to convince everyone. The impact of these proposed innovations of learning on social relations within the colony—on who could now claim to be trustworthy and who no longer could—mattered too.

Thirdly, while some historians have written about this period as one characterized by the disenchantment of the world, in which the natural and the spiritual worlds became distinct, beliefs about spiritual forces acting through the natural world altered but did not disappear. The powerful central commitment to the natural world as a text of God, which virtually no one abandoned in the eighteenth-century Atlantic world, assured that disputes would be over the nature of his messages, not the possibility of their existence. As historians of British natural philosophy have made clear, the acceptance of new understandings of the natural world were intimately entwined with shifts in religious belief and, moreover, competing understandings were weapons in the contests between competing claims to religious truth. Other historians, most notably Carolyn Merchant, have argued that in the early modern period the natural world became "disenchanted" in another sense, one in which Europeans came to experience themselves as mastering nature, the homologue of men's domination over women. In this telling, Europeans, wherever they were in the world, became alienated from nature as they reduced it simply to matter upon which they could work their will. However much some may have desired this, New Yorkers continued to experience themselves as too vulnerable to the natural world to claim such mastery. Importantly, as long as God worked his Providence through the natural world, those who believed themselves subject to him could not believe themselves so capable of completely dominating nature. Shifts in beliefs about the natural world were as much shifts in beliefs about the nature of God as about the natural world itself.

At the same time, as we have seen, Christian belief contributed the partial alienation of colonists from this particular nature. Because Christianity taught that God's nature was universal, colonists assumed a natural world that everywhere obeyed the same natural laws, whatever the variability of local forms. In this, the standardization that scientific discourses developed to deal with the plethora of natural phenomena—grids of longitude and latitude, inoculation for smallpox, explanations for lightning whether in France or Pennsylvania—affirmed this alienation from the peculiarities of the local natural world. Shielded by their technology from the need to learn the local phenomena deeply, focused on those aspects of the local natural world that lent themselves to commodity production and the market, and assured that God worked his will everywhere the same, immigrants to New Netherland and New York investigated the region's particularities only superficially. In this their knowledge was not substantially different from many others who colonized the rim of the North Atlantic or, for that matter, from those who stayed home.

As it was for those who colonized other places, as well as for ordinary people who stayed in Europe, changes in the knowledge of the natural world that were adopted by a few, or a fair number, or almost all New Yorkers were embedded in larger historical transformations. The commercial expansion of Europe, which brought more people into the colony and offered more opportunities for trade and travel for those who lived there, broke the relative isolation of the seventeenth century. Cadwallader and Jane Colden's botanical work gave them an entrée into botanical circles in Europe because Peter Collinson, their London contact, was a colonial merchant and because Isaac DuBois went to Leyden for his medical education. Young men like Samuel Spencer Skinner with his gentlemanly astronomy or Claude-Joseph Sauthier with his elegant map brought their skills and knowledge with them when they sought their fortunes in New York. The expansion of print, which moved in tandem with the commercial expansion of the Atlantic world, was similarly crucial to innovations in the knowledge of the natural world. Print allowed for the accumulation of information, for the comparisons that were the basis for the reorganization of knowledge, and for the recognition of how varied the world beyond one's personal experience was. Newspapers and almanacs brought news of changing understandings of the natural world to the literate but not necessarily learned. Because of the portability of print, coupled with the increasing circulation of people, new assertions about the true nature of the natural world circulated in patterns that followed commerce and migration.

War and politics also mattered, and both do much to explain these particular patterns of changes and indifference to change in the knowledge of nature in New York. Because New York was such a fractious colony, rhetorical claims that the promotion of new knowledge was an act of selfless devotion to the public good were shrugged off as mere window dressing. No matter how gentlemanly a decorum Charles Leslie Schaw might muster, no matter that he signed his letters to the newspaper with "F.R.S.," no one assumed that either his deportment or his fellowship in the Royal Society were at all related to his skills as a doctor. They were merely pretentious. And because New York was so fractious, the families who eventually accumulated real wealth and political power in the colony never coalesced into a unified elite, one that might use erudite knowledge of the natural world to distinguish themselves from the common sort. Practices like refined collecting of natural history or reasoned astronomy remained simply a matter of individual preference, not a widespread marker of class identity. Even if a William Livingston might write poetry about the beauties of a rural life, or promote improvements through his newspaper, the attempts by elite men to show their gentlemanly superiority through an improvement society disintegrated under the pressures of political conflict.

War and the politics of empire were the driving forces behind much of the mapping, and controversies over maps, in the colony. The Seven Years War, in particular, was a pivotal event in New York. A burst of pride in membership in a triumphal British Empire encouraged the embrace of fashions like natural philosophy demonstrations, participation in observing the transit of Venus, the establishment of a nursery called a "Linnaean Botanic Garden," and the expanded sale of gazetteers, travel narratives, and maps of the world. The war brought a number of skilled engineers and surveyors into the colony, with a consequent sharp rise in local maps based on actual measurements. But the fierce competition for the sudden seeming abundance of land in the wake of the removal of the French threat also left no doubt about whose interests were served by such seemingly objective productions like maps. And the ensuing conflicts over the sequence of British policies intended to bind the colonies more firmly within the Empire raised disquieting questions about the contrast between the cosmopolitanism represented by adopting London fashions and the virtue of old, plain ways.

As one might expect, the Revolution came hard to New York. Here it was as much a civil war as it was a war of rebellion. The New York representatives to the Continental Congresses went reluctantly, and then hesitated over declaring independence. In the summer of 1776 the British sailed into New York harbor, debarked on western Long Island, and forced George Washington and his Continental Army across the Hudson to New Jersey. For the rest of the war, until peace was finally signed in 1783, New York City was the headquarters of the British forces, and the British held Long Island. Westchester County, just north of Manhattan, was a no-man's-land in which British and Continentals faced each other and fought guerrilla style. Many people on the eastern end of Long Island suffered the British presence trucu-lently, and other Long Islanders, like the poet Jupiter Hammon with his slave master, took refuge across the sound in Connecticut, from which they raided their former neighbors who were loyal to Britain. Up the Hudson Valley and in the newly settled areas in an arc around Albany people were divided in their loyalties. Recent immigrants from Britain and discharged soldiers often stayed loyal to the Crown. Others, like tenants of great landlords, chose on the basis of those relationships, reflecting the satisfactoriness, or lack thereof, of their tenancy. Thus did many tenants of the Livingstons, who chose the Patriot cause, choose the side of the king, many tenants of Beverly Robinson, who was a loyalist, chose to rebel against him as much as against Britain, and still others, like many of the tenants of Frederick Phillipse, became Patriots with him. Many others all through New York hunkered down for the duration, avoiding choosing either side. By and large the Patriot side dominated politically in the areas north of Westchester County, successfully harassing and repressing those who chose loyalty to Britain.

The most brutal arena of warfare in the area was Iroquoia. Even though some Iroquois, like the Oneidas and Tuscaroras, chose to ally themselves with the Patriot cause, others, like the Mohawks, long most closely allied with the British and particularly loyal to the family of William Johnson, fought with the British as they had in previous wars against the French. But their choices probably didn't matter. The ferocity with which Patriot forces raided Iroquois villages, a ferocity answered by Iroquois raids on settlers' farms, reflected the widespread desire that Indians be gone altogether.

The defeat of the British altered many New Yorkers' fortunes. Samuel Spencer Skinner packed up and went back to Britain, his business ruined by the war. Cadwallader Colden died at the war's beginning but three of his children survived. Cadwallader Jr., who farmed in the Hudson Valley, was threatened by Patriot militias but managed to hang on to both his life and his property. Elizabeth Colden DeLancey saw her husband's property all forfeit for his loyalism. The youngest, David, whose frail health had necessitated such careful inoculation for smallpox, had stayed in New York City to support the British administration. He went into exile in London in 1783, seeking a way to re-establish himself while his wife and children found refuge with his brother Cadwallader. He died, however, before finding a new path. Wealthy Patriots like Philip Schuyler found their leadership less acknowledged in the wake of the war, as middling men like George Clinton rose to political power in the new republic. John Lamb, the son of the man who had first brought improved surveying and navigation instruments to New York, had been a leader of the Sons of Liberty in New York City, and after the British evacuated the city returned to political prominence. And thousands of black New Yorkers sailed away on British ships to freedom, settling in Nova Scotia, England, and the West Indies.

In the wake of the war, Iroquoia was incorporated into the new state of New York. New York now extended west as far as Lake Erie and Niagara Falls. Even Iroquois who had fought with the Patriots were reduced to living on remnants of their former territories. All of the rest was opened for settlement. And learning from its colonial experiences, the new government of the State of New York was much more careful to specify transparent surveying as the basis for land title, a process willingly embraced by land speculators and settlement companies. The hub of wheat farming shifted from the Hudson Valley north and west; farmers south of Albany adapted by moving toward the dairy farming and fruit and vegetable production that would feed a burgeoning New York City.

The embrace of improvement evident by the 1750s and 1760s grew stronger still. In 1791 the new Society for the Promotion of Agriculture, Arts, and Manufactures was established in New York City. The society relocated in 1797 to Albany when that city became the state capital, and then reorganized as the Society for the Promotion of Useful Arts in 1804. Thereafter, it was primarily concerned with promulgating innovations in agriculture. And in 1817 construction began on the Erie Canal, an improvement that would stimulate new feats in engineering. Completed in 1825, it would transform the economy of the state, allowing the western areas to thrive and making New York City the busiest port in the nation.

In the nineteenth century, in New York as elsewhere, knowledge of the natural world became less "everyday." And it did so in two senses. On the one hand, knowledge about nature came to be reorganized into the disciplines that are familiar to us: physics, biology, chemistry, geology, astronomy.

Simultaneously, these formal sciences became the province of university-trained professionals. While ordinary people might collect shells, or bird-watch, or stargaze, these were now recreations for amateurs, not arenas for the development of widely respected knowledge. At the same time, nature became deeply romanticized, increasingly associated with nationalist feeling and reverence for God. But in New York as elsewhere, only grand and wild nature qualified for such adulation. It was the extraordinary landscapes of the Hudson River and Niagara Falls that were celebrated in the art and poetry of the 1830s and 1840s and thereafter, not city pigeons, sidewalk weeds, or the gray chill of February. The former was "nature" and the latter not. Both of these—the institutionalization of formal science and the romanticism of wilderness—developed in very different political and social circumstances than those of colonial New York.

But that, as they say, is another story.

NOTES

Repositories

AAS	American Antiquarian Society
APS	American Philosophical Society
NYHS	New-York Historical Society
NYPL	New York Public Library

Collected Documents

DRCHSNY	*Documents Relative to the Colonial History of the State of New York*. Edited by E. B. O'Callaghan. 13 vols. Albany: Weed, Parsons, 1853–1887.
ERSNY	*Ecclesiastical Records, State of New York*. Compiled by Hugh Hastings. 7 vols. Albany: J. Lyon, 1901–1916.
NYHS *Collections*	New-York Historical Society, Collections

Newspapers

NYEP	*New-York Evening Post* (1744–1752)
NYG	*New-York Gazette* (1725–1744)
NYJ	*New-York Journal* (1766–1775)
NYM	*New-York Mercury* (1752–1775)
NYWJ	*New-York Weekly Journal* (1733–1751)
NYWPB	*New-York Weekly Post-Boy* (1743–1775)
RNYG	*Rivington's New-York Gazetteer* (1773–1775)

Introduction: Nature and Knowledge in a Colonial Atlantic World

1. Donald E. Stanford, "The Giant Bones of Claverack, New York, 1705," *New York History* 40 (1959): 47–61; Paul Semonin, *American Monster: How the Nation's*

First Prehistoric Creature Became a Symbol of National Identity (New York: New York University Press, 2000), 15–40.

2. For overviews of colonial New York, see Edwin G. Burrows and Mike Wallace, *Gotham: A History of New York City to 1898* (New York: Oxford University Press, 1999); Oliver A. Rink, "Before the English," and Ronald W. Howard, "The English Province," in *The Empire State: A History of New York*, ed. Milton M. Klein (Ithaca, NY: Cornell University Press, 2001).

3. The literature on the scientific revolution is enormous; a good overview with an excellent bibliography is Steven Shapin, *The Scientific Revolution* (Chicago: University of Chicago Press, 1996).

4. For examples of engagement of ordinary people, see Patricia Fara, *Sympathetic Attractions: Magnetic Practices, Beliefs, and Symbolism in Eighteenth-Century England* (Princeton: Princeton University Press, 1996); of colonial experiences affecting the development of science in Europe: Joyce E. Chaplin, *Subject Matter: Technology, the Body, and Science on the Anglo-American Fronier, 1500–1676* (Cambridge: Harvard University Press, 2001), and Richard Harry Drayton, *Nature's Government: Science, Imperial Britain, and the "Improvement" of the World* (New Haven, CT: Yale University Press, 2000); of science as inextricably related to other cultural spheres: James Delbourgo, "Electricity, Experiment and Enlightenment in Eighteenth-Century North America" (PhD diss., Columbia University, 2003); of participation of a broad range of people: Susan Scott Parrish, *American Curiosity: Cultures of Natural History in the Colonial British Atlantic World* (Chapel Hill, NC: Omohundro Institute for Early American History and Culture by University of North Carolina Press, 2006).

5. A good place to begin with this literature is Thomas F. Gieryn, *Cultural Boundaries of Science: Credibility on the Line* (Chicago: University of Chicago Press, 1999).

CHAPTER 1   *Landscape*

1. Daniel Denton, *A Brief Description of New York, Formerly Called New Netherlands*, ed. Gabriel Furman (New York: William Gowans, 1845), quotes on pp. 3–6, 14–15, 18–19.

2. Jaspar Dankers and Peter Sluyter, *Journal of a Voyage to New-York and a Tour in Several of the American Colonies in 1679–80*, trans. and ed. Henry C. Murphy (1867; reprint, Ann Arbor: University Microfilms, 1966), quotes on pp. 111, 295, 299.

3. Benjamin Schmidt, "Mapping an Empire: Cartographic and Colonial Rivalry in Seventeenth-Century Dutch and English North America," *William and Mary Quarterly*, 3rd series, 54 (1997): 549–578; John P. Snyder, *The Mapping of New Jersey: The Men and the Art* (New Brunswick, NJ: Rutgers University Press, 1973), 5–18; David Yehling Allen, *Long Island Maps and Their Makers: Five Centuries of Cartographic History* (Mattituck, NY: Amereon House, 1997), 2–12.

4. John A. Strong, *The Montaukett Indians of Eastern Long Island* (Syracuse: Syracuse University Press, 2001); Shirley W. Dunn, *The Mohicans and Their Land, 1609–1730* (Fleischmanns, NY: Purple Mountain Press, 1994); Dean R. Snow,

The Iroquois (Cambridge, UK: Blackwell, 1994); Michael Recht, "The Role of Fishing in the Iroquois Economy," *New York History* 78 (1997): 429–454.

5. Oliver A. Rink, *Holland on the Hudson: An Economic and Social History of Dutch New York* (Ithaca, NY: Cornell University Press, 1986); Thomas Eliot Norton, *The Fur Trade in Colonial New York, 1686–1776* (Madison: University of Wisconsin Press, 1974); on use of abandoned Indian fields: Alice P. Kenney, *Stubborn for Liberty: The Dutch in New York* (Syracuse, NY: Syracuse University Press, 1975), 93.

6. For villages at trailheads: Carl Nordstrom, *Frontier Elements in a Hudson River Village* (Port Washington, NY: Kennikat Press, 1973), 28; and John Reading, "Journal," New Jersey Historical Society *Proceedings* 10 (1915): 35–46, 90–110, 128–133.

7. David Steven Cohen, *The Dutch-American Farm* (New York: New York University Press, 1992); Faren R. Siminoff, *Crossing the Sound: The Rise of Atlantic American Communities in Seventeenth-Century Eastern Long Island* (New York: New York University Press, 2004); on diking, see Thomas E. Burke Jr., *Mohawk Frontier: The Dutch Community of Schenectady, New York, 1661–1710* (Ithaca, NY: Cornell University Press, 1991), 47; "Declaration of Brooklyn Officials," 24 April 1677, in *The Andros Papers, 1677–1680: Files of the Provincial Secretary of New York During the Administration of Governor Sir Edmund Andros*, ed. Peter R. Christoph and Florence A. Christoph, 3 vols. (Syracuse, NY: Syracuse University Press, 1989), II: 54–55; for the shift from furs to flour: Cathy D. Matson, *Merchants and Empire: Trading in Colonial New York* (Baltimore: Johns Hopkins University Press, 1998), 92–117.

8. On livestock observed as marked and running free, see Dankers and Sluyter, *Journal of a Voyage*, 141–144; on herders: *Huntington Town Records*, 30 May 1663, comp. Charles R. Street (Huntington NY: The Long Islander, 1887), I: 68; on fencing and yokes on hogs: *Records of the Town of East-Hampton, Long Island*, 19 May, 1651 (Sag Harbor, NY: John H. Hunt, 1887), I: 14–15; on nose rings: *Annals of Albany*, 15 Feb. 1703/04, ed. Joel Munsell, 10 vols. (Albany, NY: Joel Munsell, 1850–1869), IV: 187; Virginia DeJohn Anderson, *Creatures of Empire: How Domestic Animals Transformed Early America* (Oxford: Oxford University Press, 2004).

9. Edward Porter Alexander, ed., *The Journal of John Fontaine: An Irish Huguenot Son in Spain and Virginia, 1710–1719* (Williamsburg, VA: The Colonial Williamsburg Foundation, 1972), 113; Lewis Morris to James Alexander, 29 Jan. 1722/23, ed. Eugene R. Sheridan, *The Papers of Lewis Morris*, 3 vols. (Newark, NJ: New Jersey Historical Society, 1991–1993), I: 225–31.

10. For agricultural practices on Long Island, see 21 Dec. 1668, *Records of East-Hampton* I: 303; 8 April 1685, *Records of East-Hampton* II: 165; 16 April 1708, *Records of East-Hampton* III: 187; for Nyacks: Dankers, *Journal of a Voyage*, 124–126; for Schaghticoke: 31 March 1711, *Annals of Albany*, VI: 257; for complaint of Indians against colonists: 14 April 1676, *Andros Papers* I: 352–353; for complaint of colonists against Indians: 20 Nov. 1680, *Andros Papers* III: 444–445.

11. Adriaen van der Donck, *The Representation of New Netherland*, trans. Henry C. Murphy (New York: Bartlett and Welford, 1849), 16–17; Denton, *A Brief*

Description, 3–4; Jeremias van Rensselaer to Johan van Wely, [Oct. 1656], in *Correspondence of Jeremias van Rensselaer, 1651–1674*, trans. and ed. A.J.F. van Laer (Albany: The University of the State of New York, 1932), 34–35; Jeremias van Rensselaer to Jan van Wely, 23 Oct. 1657, in *Correspondence*, 65; on Palatine refugees eating local plants: "The Condition, Grievances, and Oppressions of the Germans in His Majesty's Province of New York in America," 20 Aug. 1720, in *ERSNY* III: 2168–2172.

12. Jacob Steendam, "Praise of New Netherland," in Christine van Boheeman, "Dutch-American Poets of the Seventeenth Century," in Rob Kroes and Henk-Otto Neuschafer, eds., *The Dutch in North America: Their Immigration and Cultural Continuity* (Amsterdam: VU University Press, 1991), 126; for an example of oystering: 14 April 1676, *Andros Papers* I: 353; for Dutch shooting game: Dankers, *Journal of a Voyage*, 131; for whaling: Strong, *Montaukett*, 48–54; for Indians selling venison, wild turkeys, and wild geese: Dankers, *Journal of a Voyage*, 123; for Indians selling oysters and fish: Charles Wooley, "A Two Years Journal in New York (1701)," in *Historic Chronicles of New Amsterdam, Colonial New York and Early Long Island*, ed. Cornell Jaray (Port Washington, NY: Ira J. Friedman, n.d.), 46; for "at an easie rate": Denton, *A Brief Description*, 2–3.

13. For "distant and strange," see Jeremias van Rensselaer to Anna van Rensselaer, 16/26 Nov. 1669, in *Correspondence*, 416; for "heathenish": Jeremias van Rensselaer to Jan Baptist van Rensselaer, 3/6 June 1660, in *Correspondence*, 225; for "slow our Conveighance": Governor Lovelace to Secretary Williamson, 3 Oct. 1670, in *DRCHSNY* III: 189–190.

14. John Miller, "A Description of the Province and City of New York (1695)" in Jaray, *Historic Chronicles*, 33; for "generation and corruption": Daniel Leeds, *An Almanack . . . 1700* (New-York: William Bradford, 1700); for "Boterberg": Dankers, *Journal of a Voyage*, 330.

15. For bounties, see *Dutch Records of Kingston, Ulster Co., NY, 1658–1684*, 9 Oct. 1663, rev. trans. by Samuel Oppenheim (n.p.: New York State Historical Association, 1912), 81; for payment to Indians: *Brookhaven Town Records*, 17 Oct. 1682, comp. Archibald C. Weeks (New York: Tobias A. Wright, 1924), 406; for wolves: Judith Maureen Adkins, "Bodies and Boundaries: Animals in Early American Experience" (PhD diss., Yale University, 1998), 111–120; the Reverend Johannes Megapolensis, "The Mohawk Indians," *Annals of Albany* IX: 133; Van der Donck, *Representation of New Netherland*, 16.

16. For understandings of climate, see H. Howard Frisinger, *The History of Meteorology to 1800* (New York: Science History Publications, 1977), 27–28; for severe winter: Jeremias van Rensselaer to Anna van Rensselaer, 8 May 1659, in *Correspondence*, 156–157; for wit: Lewis Morris, "The Looking Glasse: or, a dialogue between a barbadian and an Inhabitant of one of the northern Plantations," *Papers* I: 286–326; for "Forges and Bellows": Wooley, "Two Years Journal," 22–23; Titan Leeds, *The American Almanack . . . 1714* (New-York: William Bradford, 1714).

17. For Stuyvesant's efforts, see Evan Haefeli, "The Creation of American Religious Pluralism: Churches, Colonialism, and Conquest in the Mid-Atlantic, 1628–1688"

(PhD diss., Princeton University, 2000), 109–113; for "Bacchanalian": Reverend Hermanus Blom to the Magistrates at Wildwyck, 12 Feb. 1664, in *Dutch Records of Kingston*, 126–127; for maypole: *Andros Papers*, 8 Jan. 1678, II: 316–317.

18. For "adventurers": "Answers of Governor Andros to Enquiries about New-York," 16 April 1678, in DRCHSNY III: 260–262; for Hackensack: Dankers, *Journal of a Voyage*, 267–268.

19. For Van Ilpendam, see Donna Merwick, *Death of a Notary: Conquest and Change in Colonial New York* (Ithaca: Cornell University Press, 1999); for Melyn: Evan Haefeli, "Dutch New York and the Salem Witch Trials: Some New Evidence," in AAS *Proceedings* 110 (2000): 277–308; Joyce D. Goodfriend, *Before the Melting Pot: Society and Culture in Colonial New York City, 1664–1730* (Princeton: Princeton University Press, 1992); Philip Otterness, *Becoming German: The 1709 Palatine Migration to New York* (Ithaca, NY: Cornell University Press, 2004).

20. Simon Middleton, "How it came that the bakers bake no bread: A Struggle for Trade Privileges in Seventeenth-Century New Amsterdam," *William and Mary Quarterly*, 3rd series, 58 (2001): 345–370; Matson, *Merchants and Empire;* Alan Tully, *Forming American Politics: Ideals, Interests, and Institutions in Colonial New York and Pennsylvania* (Baltimore: Johns Hopkins University Press, 1994), 11–27, 49–68.

21. Lawrence H. Leder, *Robert Livingston, 1654–1728, and the Politics of Colonial New York* (Chapel Hill: Published for the Institute for Early American History and Culture by University of North Carolina Press, 1961); for "as Christian and Turk": Governor Fletcher to the Committee of Trade, 10 Nov. 1693, in *DRCHSNY* IV: 72–73.

22. Dennis J. Maika, "Slavery, Race, and Culture in Early New York," *de Halve Maen* 73 (2000): 27–33; Leslie M. Harris, *In the Shadow of Slavery: African Americans in New York City, 1626–1836* (Chicago: University of Chicago Press, 2003), 11–47; for Jacob: 2 Sept. 1679, *Andros Papers* III: 140–141.

23. Thelma Wills Foote, "'Some Hard Usage': The New York City Slave Revolt of 1712," *New York Folklore* 18 (2000): 147–160; for "ye Nations of": John Sharpe, "The Negro Plot of 1712," *New York Genealogical and Biographical Record* 21 (1890): 162–163.

24. Michael Kammen, *Colonial New York: A History* (New York: Oxford University Press, 1975), 31–44, 75–78.

25. Peter Fauconnier, in his survey book for Staten Island, April-June 1723, Staten Island Box, NYHS; for most grants recorded without surveys: "Mr. Lewin's Report on the Government of New-York," 4 May 1680, in *DRCHSNY* III: 302–308; compare William Roome, *The Early Days and Early Surveys of East New Jersey* (Morristown, NJ: Jerseyman Press, 1883); Roger J. P. Kain and Elizabeth Baigent, *The Cadastral Map in the Service of the State: A History of Property Mapping* (Chicago: University of Chicago Press, 1992); Patricia Fara, *Sympathetic Attractions: Magnetic Practices, Beliefs, and Symbolism in Eighteenth-Century England* (Princeton: Princeton University Press, 1996).

26. For "rather increased them," see Caparus van Zuuren to the Classis of Amsterdam, 25 June 1681, in *ERSNY* II: 771–777; for example of surveyors who signed with marks: "Certificate for land claimed by Ryck Leydecker," 30 Jan. 1676/77,

in *Andros Papers* II: 17; for "lump by miles": "Report of the Board of Trade," 19 Oct. 1698, in *ERSNY* II: 1244–1245.

27. For "by the stump," see Fauconnier's survey book for Staten Island, n.p.; for "malice and jealousy": Peter Fauconnier, 1716 boundary papers, Fauconnier Papers, NYHS; for "three Neighbours": Armand La Potin, "The Minisink Grant: Partnerships, Patents, and Processing Fees in Eighteenth-Century New York," *New York History* 56 (1975): 39.

28. For process and cost, see La Potin, "The Minisink Grant"; Warrents of Survey, Powers of Attorney, Indian Deeds, Misc. Records, 1721–1776, 1 microfilm reel, New York State Archives; for Esopus: 27 April 1677, *Andros Papers* II: 57–59; for Iroquois skill in retaining control of territory: Wyllys Terry, "Negotiating the Frontier: Land Patenting in Colonial New York" (PhD diss., Boston University, 1997); for Mohicans: Dunn, *Mohicans*, 136–137.

29. For Livingston's patent : Leder, *Robert Livingston*, 32–35.

30. Sung Bok Kim, *Landlord and Tenant in Colonial New York: Manorial Society, 1664–1775* (Chapel Hill: University of North Carolina Press, 1978), 3–128.

31. For "several most extravagant," see Earl of Bellomont to the Lords of Trade, 21 Oct. 1698, in *DRCHSNY* IV: 397–401; for maps generally: Sara Stidstone Gronim, "Geography and Persuasion: Maps in British Colonial New York," *William and Mary Quarterly*, 3rd series, 58 (2001): 373–402.

32. For "this whole Province," see Earl of Bellomont to the Lords of Trade, 21 Oct. 1698, in *DRCHSNY* IV: 397–401.

33. For "breadth we know not," see "Report of the Board of Trade on the Affairs of the Province of New-York," in *ERSNY* II: 1244–1245; for voiding two patents: Georgiana C. Nammack, *Fraud, Politics, and the Dispossession of the Indians: The Iroquois Land Frontier in the Colonial Period* (Norman: University of Oklahoma Press, 1969), 13–15; for 2,000 acre restriction: "Additional Instructions for Lord Lovelace," 20 July 1708, in *DRCHSNY* V: 54–55.

34. Daniel K. Richter, *The Ordeal of the Longhouse: The Peoples of the Iroquois League in the Era of European Colonization* (Chapel Hill: Published for the IEAHC by University of North Carolina Press, 1992), 50–190.

35. Gronim, "Geography and Persuasion," 378–379.

36. For accounts of surveying the NY-NJ boundary, see James Alexander, "Journal Kept by James Alexander, Surveyor-General of Newjersey," in James Alexander Papers, NYHS; Reading, "Journal"; for "agreed on all hands": Colonel Schuyler to the Lords of Trade, 31 Oct. 1719, in *DRCHSNY* V: 531–532; Philip J. Schwarz, *The Jarring Interests: New York's Boundary Makers, 1664–1776* (Albany: SUNY Press, 1979), 75, 94–95.

37. For use of Visscher map: Snyder, *Mapping of New Jersey*, 14; for boundary determination: "Warrant to Prepare a Patent for Sir George Carteret," 23 July 1674, in *DRCHSNY* III: 223–224.

38. Alexander, "Journal"; Reading, "Journal."

39. For Jarrett's reversal, see "Examination of Allane Jarrett," in East Jersey Proprietors, "The Memorial of the Council of the Proprietors of the Eastern Division of New-Jersey, 1753," Special Collections, Alexander Library, Rutgers University; for "Art of Man": John Barclay to Lewis Morris, 12 Oct. 1719, in *Morris Papers*

I: 195; for "without offering": Peter Schuyler to Lords of Trade, 31 Oct. 1719, in *DRCHSNY* V: 531–32.

40. For "map immersed," see Denis Wood with John Fels, *The Power of Maps* (New York: Guildford Press, 1992), 34; for Schuyler: "Journal of Captain Arent Schuyler," 3 Feb. 1693/94, in *DRCHSNY* IV: 98–99.

41. For the census, see Evarts B. Greene and Virginia D. Harrington, *American Population Before the Federal Census of 1790* (New York: Columbia University Press, 1932), 90; for "divers Coullers": Sarah Kemble Knight, *The Journal of Madam Knight* (1920; reprint, Upper Saddle River, NJ: Gregg Press, 1970), 52–56.

CHAPTER 2 *Body and World*

1. Robert Livingston Jr. to Philip Livingston, 30 May 1716, in Robert Livingston Papers, The Morgan Library, New York City, reel 6; Alida Livingston to Robert Livingston, 12 Oct. 1717, in Alida Livingston Papers, transcripts, The Morgan Library, New York City, n.p.; Johanna Livingston to Robert Livingston, 19 Nov. 1712 and 19 Jan. 1712/13, in Robert Livingston Papers, reel 3; for Livingstons: Linda Biemer, ed., "Business Letters of Alida Schuyler Livingston, 1680–1726," *New-York History* 63 (1982): 184–207.

2. For Dutch medical self-help books, see A. T. van Deursen, *Plain Lives in a Golden Age: Popular Culture, Religion and Society in Seventeenth Century Holland*, trans. Maarten Ultee (Cambridge: Cambridge University Press, 1991), 238; for the same in England: Carey Balaban, Jonathon Erlen, and Richard Siderits, "Introduction" to *The Skilful Physician* (1656; reprint, Amsterdam: Harwood Academic Publishers, 1997), ix–xxiv; for Van Imbrock's estate: David William Voorhees, "Editor's Corner," *de Halve Maen* 73 (2000): 70; for Bellomont and ministers: Austin Baxter Keep, *The Library in Colonial New York* (New York: The DeVinne Press, 1909), 8–65.

3. Leeds' first almanac in New York: *An Almanack For the Year of Christian Account 1694* (New-York: William Bradford, 1694); for English circulation of almanacs: Bernard Capp, *Astrology and the Popular Press: English Almanacs, 1500–1800* (London: Faber and Faber, 1979), 23; for the Livingstons' order: Robert Livingston to William Bradford, 3 Nov. 1722, in Robert Livingston Papers, reel 4; for selling on Long Island: Titan Leeds, *The American Almanack . . .* 1729 (Philadelphia: S. Keimer, n.d.); *The Husband-man's Guide, in Four Parts*, 2nd ed. (New-York: William and Andrew Bradford, 1712).

4. George Keith, "Truth Advanced in the Correction of many Gross & hurtful Errors" ([New York: William Bradford], 1694), 119; Anna van Rensselaer to Jeremias van Rensselaer, 26 Dec. 1654, in *Correspondence of Jeremias van Rensselaer, 1651–1674*, trans. and ed. A.J.F. van Laer (Albany: The University of the State of New York, 1932), 14–16; on the development of the idea of humors and the relationship to treatment: Carole Rawcliffe, *Medicine and Society in Later Medieval England* (Phoenix Mill, Eng.: Alan Sutton Publishing, 1995), 29–57; on the preeminence of blood by the 17th century: Trudy Ann Eden, "'Makes Like, Makes Unlike': Food, Health and Identity in the Colonial Chesapeake" (PhD diss., Johns Hopkins University, 1999), 79.

5. On beliefs about differences between men and women: Lorraine Daston, "The Naturalized Female Intellect," *Science in Context* 5 (1992): 217–218; for the origins of ideas that linked temperament with bodily characteristics: Nancy G. Siraisi, *Medieval and Early Renaissance Medicine: An Introduction to Knowledge and Practice* (Chicago: University of Chicago Press, 1990), 101–104.

6. Robert Livingston, "Journal kept by Robert Livingston, 1695," in Robert Livingston Papers, reel 1.

7. Jeremias van Rensselaer to Anna van Rensselaer, 21 Oct. 1664, *Correspondence*, 367–368; Bradford's first advertisement was for Lockyers Universal Pill: Daniel Leeds, *An Almanack . . . 1698* (New-York: William Bradford, 1698); for example of order from Boston: Oloff Stevensson van Cortlandt to Maria van Rensselaer, [16 Jan. 1678], in *Correspondence of Maria van Rensselaer*, trans. and ed. A.J.F. van Laer (Albany: The University of the State of New York, 1935), 19–20; two page list of medicines ordered, 1724, in Robert Livingston Papers, reel 4; for seeds and plants: Peter Stuyvesant to the Directors, 17 Sept., 1659, in *ERSNY* I: 451; Directors to Peter Stuyvesant, 16 April 1660, in *ERSNY* I: 474; for almanac recipe: Daniel Leeds, *The American Almanack . . . 1705* (New-York: William Bradford, 1705).

8. [Adriaen van der Donck], *The Representation of New Netherland, Concerning Its Location, Productiveness and Poor Condition*, trans. Henry C. Murphy (New York: Bartlett and Welford, 1849), 16–17; for Indian corn: Daniel Leeds, *An Almanack . . . 1695* ([New-York]: William Bradford, 1694); for tobacco: Daniel Leeds, *The American Almanack . . . 1713* (New York: William Bradford, 1713).

9. For the genealogy of medical ideas, see Frederick Sargent II, *Hippocratic Heritage: A History of Ideas About Weather and Human Health* (New York: Pergamon Press, 1982); Roy Porter, *The Greatest Benefit to Mankind: A Medical History of Humanity* (New York: W.W. Norton, 1998), 201–211; Daniel Leeds, *The American Almanack . . . 1708* (New-York: William Bradford, 1708); for "moderately true" quote and human fallibility in medicine: Daniel Leeds, *An Almanack . . . 1695*.

10. For Sloughter, see Johannes van Geisen, et al., to the Classis of Amsterdam, 21 Oct. 1698, in *ERSNY* II: 1246–1261; Robert Livingston, "Journal kept by . . . 1695"; Jaspar Dankers and Peter Sluyter, *Journal of a Voyage to New York and a Tour in Several of the American Colonies in 1679–80*, trans. and ed., Henry C. Murphy (1867; reprint Ann Arbor: University Microfilms, 1967), 110; Titan Leeds, *The American Almanack . . . 1715* (New-York: William Bradford, 1715); for mustard seed: Daniel Leeds, *The American Almanack . . . 1713*; for beliefs about which foods affected which humors: Eden, "Makes Like, Makes Unlike," 79–86.

11. Titan Leeds, *Almanack . . . 1715*; for the origin of ideas about the relationship between seasons and constitutions: Sargent, *Hippocratic Heritage*, 46–62; for Mary Livingston: G. Saltonstall to Robert Livingston, 3 Sept. 1712, in Robert Livingston Papers, reel 3; Deposition of James Leigh, 15 Feb. 1713/14, in *ERSNY* III: 2022.

12. Titan Leeds, *Almanack . . . 1715*.

13. Charles Wooley, "A Two Years Journal in New York (1701)," in *Historic Chronicles of New Amsterdam, Colonial New York and Early Long Island*, ed. Cornell Jaray (Port Washington, NY: Ira J. Friedman, n.d.), 28–29, 45.

14 For "Winter and Summer"· William Hyde, "The Hyde Manuscript· Captain William Hyde's Observations of the 5 Nations of Indians at New Yorke, 1698," ed. William N. Fenton, *American Scene* 6 (1965): n.p.; for "fiddle-faddle": Johannes Megapolensis, "The Mohawk Indians," in *Annals of Albany*, ed. Joel Munsell, 10 vols. (Albany: Joel Munsell, 1850–1869), IX: 138–139; for "Rules for Sweating," *Husband-man's Guide*, 30.

15. For ruling humor quote, see Daniel Leeds, *An Almanack . . . 1698* ; for poem: Daniel Leeds, *The American Almanack . . . 1709* (New-York: William Bradford, 1709); for absence in early New England almanacs, see Peter Eisenstadt, "The Weather and Weather Forecasting in Colonial America" (PhD diss., New York University, 1990), 87–98; for the widespread acceptance in later New England almanacs by the late seventeenth century in New England: Michael P. Winship, *Seers of God: Puritan Providentialism in the Restoration and Early Enlightenment* (Baltimore: Johns Hopkins University Press, 1996), 112; for the origins : Rex Jones, "Genealogy of a Classic: The English Physitian of Nicholas Culpepper" (PhD diss., University of California, San Francisco, 1984), 218–219.

16. For "Golden Nails" quote, see B.A., *An Astronomical Diary, or an Almanack . . . 1723* (New-York and Philadelphia: William and Andrew Bradford, n.d.); for Ptolemaic cosmology in colonial almanacs: J. Rixey Ruffin, "'Urania's Dusky Vails': Heliocentrism in Colonial Almanacs, 1700–1735," *New England Quarterly* 70 (1997), 306–313; for astrology's effects as "obvious": Keith, "Truth Advanced," 23–24.

17. For eclipse prediction, see Daniel Leeds, *An Almanack . . . 1699* (New-York: William Bradford, 1699); for "Useful to planters": Daniel Leeds, *An Almanack . . . 1694*; for Algoquian women planting: Lynn Ceci, "Watchers of the Pleiades: Ethnoastronomy among Native Cultivators in Northeastern North America," *Ethnohistory* 25 (1978), 305; for Sanders anecdote: Charles T. Gehring and Robert S. Grumet, "Observations of the Indians from Jasper Danckaerts's Journal, 1679–1680," *William and Mary Quarterly*, 3rd series, 44 (1987), 109–110; for Leeds's scorn: Daniel Leeds, *The American Almanack . . . 1707* ([New-York]: William Bradford, 1707).

18. For the English reforms, see Patrick Curry, *Prophecy and Power: Astrology in Early Modern England* (Princeton: Princeton University Press, 1989) , 64–68, 78–83; nonetheless, it remained a complex body of knowledge: Ann Geneva, *Astrology and the Seventeenth Century Mind: William Lily and the Language of the Stars* (Manchester, UK: Manchester University Press, 1995); for Daniel Leeds's birth: Titan Leeds, *The American Almanack . . . 1722* (New-York: William Bradford, 1722); for weather chart: Daniel Leeds, *An Almanack . . . 1699*; for "Bundle of Experience": Titan Leeds, *The American Almanack . . . 1729.*

19. For Daniel Leeds's reliance on ephemerides, see Daniel Leeds, *An Almanack . . . 1700* (New-York: William Bradford, 1700); Titan Leeds, *The American Almanack . . . 1714* (New-York: William Bradford, 1714); for this reliance among almanac makers generally, see John T. Kelly, *Practical Astronomy During the*

Seventeenth Century: Almanac Makers in America and England (New York: Garland Publishing, 1991), 112–113; for scorn for Philadelphian: Daniel Leeds, *The American Almanack . . . 1708.*

20. For Clapp's almanac: John Clapp, *An Almanack for the Year 1697* (New-York: William Bradford, 1697); for Clapp: *New-York Genealogical and Biographical Record* 112 (1981): 135–136.

21. For the origin of the "Man of Signs," see W.S.C. Copeman, *Doctors and Disease in Tudor Times* (London: Dawson's of Pall Mall, 1960), 103–105; the first appearance of the woodcut in New York: Daniel Leeds, *The American Almanack . . . 1705*; for "extream perilous": *Husband-man's Guide*, 6–7.

22. For influence of the heavens on farming, see *Husband-man's Guide*, 3, 7–8, 14; similar advice was scattered throughout the Leeds's almanacs; for "Opinion of Astrological": Daniel Leeds, *An Almanack . . . 1697* (New-York: William Bradford, 1697); Daniel Leeds gave dates for the optimal gathering of herbs in his almanacs for 1694, 1695, 1696, 1697, 1698, and 1699; Titan Leeds gave them for 1715, 1718, 1729, and 1730; for marginalia: Daniel Leeds, *The American Almanack . . . 1705*, photostat copy in Rare Books and Manuscripts, NYPL; Philip Livingston to Robert Livingston, 4 Jan. 1723/24, in Robert Livingston Papers, reel 4.

23. Earth "moist and open": Daniel Leeds, *An Almanack . . . 1695*; for a similar characterization: *Husband-man's Guide*, 3; for "English Grain": Daniel Denton, *A Brief Description of New York, Formerly called New Netherlands*, ed. Gabriel Furman (1670; reprint, New York: William Gowans, 1845), 3–4; for "grows lean": Daniel Leeds, *An Almanack . . . 1700.*

24. Eugene R. Sheridan, *Lewis Morris, 1671–1746: A Study in Early American Politics* (Syracuse, NY: Syracuse University Press, 1981); poem dated 1729: Lewis Morris, *Papers of Lewis Morris*, ed. Eugene R. Sheridan, 3 vols. (Newark: New Jersey Historical Society, 1991–1993), I: 366–372; on soil: Lewis Morris to John Morris, 22 April 1730, in *Papers*, I: 377–381; on night air: Lewis Morris to Robert Hunter Morris, 4 April 1733, in *Papers*, II: 32–33.

25. For Lord Cornbury, see Patricia U. Bonomi, *The Lord Cornbury Scandal: The Politics of Reputation in British America* (Chapel Hill: University of North Carolina Press, 1998); for Megapolensis: Willem Frijhoff, "New Views on the Dutch Period of New York," *de Halve Maen* 71 (1998), 34; for Cortelyou: Brenda Safer, "Jacques Corteljou, The Cartesian: The Connection Between René Descartes and Jacques Corteljou," *de Halve Maen* 73 (2000), 71–76; for Dankers's description: Dankers, *Journal of a Voyage*, 127–128; Ronald W. Howard, "Education and Ethnicity in Colonial New York" (PhD diss., University of Tennessee, 1978).

26. For healers in the Netherlands, see van Deursen, *Plain Lives*, 237–238; for Kerfbyle: Earl of Bellomont to the Lords of Trade [n.d. but with 1698 papers], in *DRCHSNY* IV: 712–726.

27. For reference to midwife, see Jeremias van Rensselaer to Oloff Stevensen van Cortlandt, Jan. 5/15 1671, in *Correspondence*, 432–433; Harmen Meyndertsz van den Bogaert, *A Journey into Mohawk and Oneida Country, 1634–1635*, ed. and trans. Charles T. Gehring and William A. Starna (Syracuse, NY: Syracuse University Press, 1988), 4; for de Hinse: Jeremias van Rensselaer to Oloff van Cortlandt, 23 Nov./3 Dec. 1665, in *Correspondence*, 384–385; for Provoost: Edwin R.

Purple, *Genealogical Notes of the Provoost Family* (New York: privately printed, 1875), 24; for Morris's rhyme: "The Mock Monarchy," 1729, *Morris Papers* I: 366–377.

28. For Morris's doctor: John Johnston to Lewis Morris, 10 Aug. 1709, in *Morris Papers* I: 102–103; as apothecary-trained: editor's footnote, *DRCHSNY* V: 34; for Peter the doctor: Kenneth Scott, "The Slave Insurrection in 1712," *New-York History* 45 (1961): 51–58; for jurymen: 21 May 1670 in *Records of the Town of East-Hampton, Long Island*, 9 vols. (Sag Harbor, NY: John H. Hunt, 1887–[1958]), I: 328; 30 April 1700 in *Annals of Albany*, 4 vols. ed. Joel Munsell (Albany, NY: J. Munsell, 1850–1869), IV: 109–110; for Sloughter's death: Commander-in-Chief and Council of New-York to Mr. Blathwayt, 6 Aug. 1691, in *DRCHSNY* III: 794–796.

29. For "the staggers," see Dankers, *Journal of a Voyage*, 264; for veterinary medical advice generally: *The Husband-man's Guide*, Part III; for Cortelyou's life and surveying: John van Zandt Cortelyou, *The Cortelyou Genealogy* (Lincoln, NE: Press of Brown Printing Service, 1942), 7–64.

30. For Kerfbyle on Govenor's Council: Earl of Bellomont to the Lords of Trade [n.d. but with 1698 papers], in *DRCHSNY* IV: 712–726; for church elder: Evan Haefili, "Dutch New York and the Salem Witchtrials: Some New Evidence," AAS *Proceedings* 110 (2000): 278; for Kerfbyle's net worth: Heads of Accusation against the Earl of Bellomont, 11 March 1700, in *DRCHSNY* IV: 620–623.

31. For "good Old Woman," see Daniel Leeds, *An Almanack . . . 1695.*

32. For political theory, see Jean Bethke Elshtain, *Public Man, Private Woman: Women in Social and Political Thought* (Princeton: Princeton University Press, 1981), 102–127; for family economy: Susan Dwyer Amussen, *An Ordered Society: Gender and Class in Early Modern England* (Oxford: Basil Blackwell, 1988); for association with "nature": Carolyn Merchant, *The Death of Nature: Women, Ecology, and the Scientific Revolution* (San Francisco: Harper and Row, 1980), 127–148; Mary Fissell, "Gender and Generation: Representing Reproduction in Early Modern England," *Gender and History* 7 (1995): 433–456; poem: Daniel Leeds, *The American Almanack . . . 1708.*

33. Joan R. Gundersen and Gwen Victor Gampel, "Married Women's Legal Status in Eighteenth-Century New York and Virginia," *William and Mary Quarterly*, 3rd series, 39 (1982): 114–134; David E. Narrett, *Inheritance and Family Life in Colonial New York City* (Ithaca, NY: Cornell University Press, 1992).

34. For example of men nursing neighbors, see two entries for 17 Aug. 1704, *Records . . . East-Hampton* III: 96–97, 98–100; for midwives: Adriana E. van Zwieten, "'[O]n her woman's troth': Tolerance, Custom, and the Women of New Netherland," *de Halve Maen* 72 (1999): 10–11.

35. For "beyond his Skill," see Daniel Leeds, *An Almanack . . . 1698*; for "exact enough": Daniel Leeds, *An Almanack . . . 1694.*

CHAPTER 3   *Anomalies*

1. Deposition of Samuel Parsons, 19 Feb. 1657, *Records of the Town of East-Hampton, Long Island* (Sag Harbor, NY: John H. Hunt, 1887) I: 128–130.

2. Randall Balmer, *A Perfect Babel of Confusion: Dutch Religion and English Culture in the Middle Colonies* (New York: Oxford University Press, 1989); Richard W. Pointer, *Protestant Pluralism and the New York Experience: A Study of Eighteenth-Century Religious Diversity* (Bloomington: Indiana University Press, 1988).

3. Reverend J. Megapolensis to the Classis of Amsterdam, 14 Aug. 1657, in *ERSNY* I: 399–400; [Daniel Leeds], "News of a Strumpet" ([New-York: William Bradford], 1701); Richard van Rensselaer to Jeremias van Rensselaer, 30 Nov. 1658, in *Correspondence of Jeremias van Rensselaer, 1651–1674*, trans. and ed. A.J.F. van Laer (Albany: The University of the State of New York, 1932), 114–121; Balmer, *Babel of Confusion*, 16–20.

4. For Cornbury: Ronald W. Howard, "The English Province," in *The Empire State: A History of New York*, ed. Milton M. Klein (Ithaca, NY: Cornell University Press, 2001), 136–138; Patricia U. Bonomi, *The Lord Cornbury Scandal: The Politics of Reputation in British America* (Chapel Hill: For Omohundro Institute for Early American History and Culture by University of North Carolina Press, 1998), 70–72; for "profane Lampoons": Address, 19 Feb. 1713/1714, in *ERSNY* III: 2025–2026.

5. Dennis J. Maika, "Slavery, Race, and Culture in Early New York," *de Halve Maen* 73 (2000): 30; for "Spanish Indians": John Sharpe, "The Negro Plot of 1712," *New York Genealogical and Biographical Record* 21 (1890): 162–163; for Neau: Graham Russell Hodges, *Root and Branch: African Americans in New York and East Jersey, 1613–1863* (Chapel Hill: University of North Carolina Press, 1999), 55–63; for 1640s Mohawk baptisms: Oliver A. Rink, "Before the English," in Klein, *Empire State*, 79–80; for other efforts: Daniel K. Richter, "'Some of them . . . Would Always Have a Minister with Them': Mohawk Protestantism, 1683–1719," *American Indian Quarterly* 16 (1992): 471–484.

6. Depositions against Justice Whitehead, 3 Feb. 1702/03, in *ERSNY*: III, 1516.

7. David William Voorhees, "The 'fervent Zeale' of Jacob Leisler," *William and Mary Quarterly*, 3rd series, 51 (1994): 447–472; Firth Haring Fabend, "'According to Holland Custome': Jacob Leisler and the Loockermans Estate Feud," *de Halve Maen* 67 (1994), 1–8; for "begat by an Incubus": Balmer, *Perfect Babel*, 41; for "persecuted two true Protestants": Evan Haefeli, "The Creation of American Religious Pluralism: Churches, Colonialism, and Conquest in the Mid-Atlantic, 1628–1688" (PhD diss., Princeton University, 2000), 334; see also Evan Haefeli, "Leislerians in Boston: Some Rare Dutch Colonial Correspondence," *de Halve Maen* 73 (2000): 77–81.

8. For "rains every day," see Alida Livingston to Robert Livingston, 24 Aug. 1720, Papers of Alida Livingston, typescript, The Pierpont Morgan Library; for "stage of Action": William Vesey, *A Sermon Preached in Trinity-Church in New-York in America, May 12, 1709* (New-York: William Bradford, 1709); for "perished in my sorrow": Maria van Rensselaer to Richard van Rensselaer, 12 Nov. 1684, in *Correspondence of Maria van Rensselaer, 1669–1689*, trans. and ed. A.J.F. van Laer (Albany: The University of the State of New York, 1935), 168–175; for Providence: Alexandra Walsham, *Providence in Early Modern England* (Oxford: Oxford University Press, 1999).

9. For example of attributing illness to postlapsarian sinfulness, see George Keith, *Truth Advanced in the Correction of many Gross & hurtful Errors* ([New York: William Bradford], 1694), 20–22; for "vain is the help of man": John Nicoll to Robert Livingston, 12 Aug. 1795, Robert Livingston Papers, Morgan Library, reel 4; for "Agues shake": Titan Leeds, *The American Almanack . . . 1715* (New-York: William Bradford, 1715); for Jaspar Dankers: Jaspar Dankers and Peter Sluyter, *Journal of a Voyage to New York and a Tour in Several of the American Colonies in 1679–80*, trans. and ed. Henry C. Murphy (1867; reprint, Ann Arbor, MI: University Microfilms, 1966), 317 ; for "prepared for that great Change": Philip Livingston to Robert Livingston, 25 Aug. 1725, Robert Livingston Papers, reel 4.

10. Daniel Leeds, *An Almanack . . . 1700* (New-York: William Bradford, 1700); for fire: Captain Leisler to King William and Queen Mary, 20 Aug. 1689, in *ERSNY* II: 972–973.

11. For "Not serving God," see Anna van Rensselaer to Jeremias van Rensselaer, 19 Feb. 1659, in *Correspondence*, 130–132; for storm: Robert Livingston, "Journal kept by Robert Livingston During His Voyage to England, Dec. 9, 1694–Oct. 3 1695," trans. A.J.F. van Laer, in Robert Livingston Papers, reel 1; for fast day: [Peter Stuyvesant], "Order Appointing a Day of General Fasting and Prayer for Oct. 15, 1659," 30 Sept. 1659, in *ERSNY* I: 451–453; for earthquake: William Corbin, *A Sermon Preached at Kings Town in Jamaica upon the 7th of June, Being the Anniversary FAST for that Dreadful Earth-Quake which happened there in the Year 1692* (New-York: William Bradford, 1703).

12. For "unfailing cure," see [Adriaen van der Donck], *The Representation of New Netherland, Concerning its Location, Productiveness and Poor Condition*, trans. Henry C. Murphy (New York: Bartlett and Welford, 1849), 16 ; for "Penny-royal": Charles Wooley, "A Two Years Journal in New York and Part of Its Territories in America" (London, 1701), in *Historic Chronicles of New Amsterdam, Colonial New York and Early Long Island*, ed. Cornel Jaray (Port Washington, NY: Ira J. Friedman, n.d.), 42–43; for the idea of local cures for local ailments: Rex Jones, "Geneology of a Classic: The English Physitian of Nicholas Culpepper" (PhD diss., University of California at San Francisco, 1984), 211–216; Daniel Leeds, *An Almanack . . . 1695* ([New York]: William Bradford, 1694.).

13. For "showed us comets," see Jeremias van Rensselaer to Jan Baptist van Rensselaer, 5/15 April, 1665, in *Correspondence*, 374–377; for "ye Dreadfull Comett": Commissaries of Albany to Captain Brockholes, 1 Jan. 1681, in *ERSNY* II: 758; Anthony Brockholes to the Commissaries of Albany, 13 Jan. 1681, in *ERSNY* II: 759; Sara Schechner Genuth, *Comets, Popular Culture, and the Birth of Modern Cosmology* (Princeton: Princeton University Press, 1997), 17–88.

14. For "increases daily," see Ensign Nyssen to Governor Stuyvesant, 11 Jan. 1664, in *DRCHSNY* XIII: 355–356; for smallpox as plague: Governor Hunter to the Lords of Trade, 12 Nov. 1716, in *DRCHSNY* V: 481; for regulations for cleanliness: Albany Common Council, 6 Nov. 1686, *Annals of Albany* II: 84; for forbidding travel: Proclamation of 2 Sept. 1702, *Annals of Albany* IV: 160–161; for the providential significance of epidemics: Walsham, *Providence*, 156–166.

15. For "a decaying people," see George Muirson to the Secretary of the SPG, 9 Jan. 1707/08, in *ERSNY* III: 1695–1697; for Indian epidemics: John Duffy, *Epidemics*

in Colonial New York (Baton Rouge: Louisiana State University Press, 1953), 70–79.

16. For "Hand of God," see Daniel Denton, *A Brief Description of New York, Formerly Called New Netherlands*, ed. Gabriel Furman (1670; reprint, New York: William Gowans, 1845), 6–7; for "the Last Day": Van der Donck, *Representation*, 36–37; for smallpox advice: His Excellency Brigadier Hunter's reply to the 5 Nations, 13 June 1717, in *DRCHSNY* V: 485–486; David Shumway Jones, *Rationalizing Epidemics: Meanings and Uses of American Indian Mortality since 1600* (Cambridge: Harvard University Press, 2004), 21–67.

17. For Dekanissore, see 13 June 1717, in *DRCHSNY* V: 486–487; for Iroquois beliefs: James W. Herrick, *Iroquois Medical Botany*, ed. Dean R. Snow (Syracuse: Syracuse University Press, 1995), 15, 25–27, 34.

18. Daniel K. Richter, "'Minister with Them,'" 479; for "a clumsy vizard": John Bartram, *Observations on the Inhabitants, Climate, Soil, Rivers, Productions, Animals, and Other Matters Worthy of Notice* (1751; reprint, New York: Arno Press, 1974), 43; for False Faces: William N. Fenton, *The False Faces of the Iroquois* (Norman: University of Oklahoma Press, 1987), with observations on changes in the masks by George Hamell on 71–72.

19. For "falling in Gemini," see Daniel Leeds, *The American Almanack . . . 1704* (New-York: William Bradford, 1704); for "Second cause": Daniel Leeds, *An Almanack . . . 1697* (New-York: William Bradford, 1697); for alignment with providentialism: Bernard Capp, *Astrology and the Popular Press: English Almanacs 1500–1800* (London: Faber and Faber, 1979), 73–97, 131–144; for techniques of interpretation: Herbert Leventhal, *In the Shadow of the Enlightenment: Occultism and Renaissance Science in Eighteenth-Century America* (New York: New York University Press, 1976), 13–65.

20. For "not for Man to be positive," see Titan Leeds, *The American Almanack . . . 1722* (New-York: William Bradford, 1722); for rhyme: Daniel Leeds, *The American Almanack . . . 1704.*

21. For "Wrath of Great Men," see Titan Leeds, *The American Almanack . . . 1725* (New-York: William Bradford, 1724); for "Clergy Men": Titan Leeds, *The American Almanack . . . 1722* (New-York: William Bradford, 1721); for denial that anyone in particular was meant: Titan Leeds, *The American Almanack . . . 1719* (New-York: William Bradford, n.d.); for Philadelphia natal chart prediction: Daniel Leeds, *The American Almanack . . . 1712* (New-York: William and Andrew Bradford, 1712).

22. For "the hands of the Lord," see Stephanus van Cortlandt to Maria van Rensselaer, [Jan.] 1684, *Correspondence of Maria van Rensselaer*, 138–139.

23. For Gabrill Legett, see Westchester County Historical Society, *Publications*, 8 vols. (White Plains, NY: n.p, 1924–30), II: 87–88; for man tempted by the Devil: Dankers, "Journey to New York," 287–293; for "the same everywhere": J. Megapolensis to S. Drusius, 14 Aug. 1657, in *ERSNY* I: 399–400.

24. The literature on witchcraft is enormous; for a good overview, see Ian Bostridge, *Witchcraft and Its Transformation, c. 1650–c. 1750* (Oxford: Clarendon Press, 1997); for legal regimes in England and its colonies: Leventhal, *Shadow of the Enlightenment*, 78–81; for Dutch skepticism: Evan Haefili, "Dutch New York

and the Salem Witch Trials: Some New Evidence," AAS, *Proceedings* 110 (2000): 290.

25. For Garlick, see *Records . . . East-Hampton* I: 128–155.

26. For Wright and Harrison: "Witchcraft in New York," NYHS *Collections*, 2 (1869): 273–276; for the Hall case: E. B. O'Callaghan, *The Documentary History of the State of New York*, 4 vols. (Albany: Weed, Parsons and Co., 1849–1851), IV: 133–138.

27. For Mather's inquiry, see "Witchcraft in New York," 276; see also Haefili, "Dutch New York and the Salem Witch Trials"; for poppet: Robert Blair St. George, *Conversing by Signs: Poetics of Implication in Colonial New England Culture* (Chapel Hill: University of North Carolina Press, 1998), 188.

28. For Henry Smith, see 30 Nov. 1678, in *The Andros Papers, 1677–1680: Files of the Provincial Secretary of New York During the Administration of Governor Sir Edmund Andros*, ed. Peter R. Christoph and Florence A. Christoph, 3 vols. (Syracuse, NY: Syracuse University Press, 1989), III: 30–31; for William Taylor: 8 June, 1680, *Andros Papers* III: 290–292; for "witches and devills": 9 Sept. 1680, *Andros Papers* III: 386–387; for "you are a sleepe": Deposition of Mrs. Gardiner, 24 Feb. 1657, in *Records . . . East-Hampton* I: 133; for the power of speech: Jane Kamensky, "Words, Witches, and Women Trouble: Witchcraft, Disorderly Speech, and Gender Boundaries in Puritan New England," *Essex Institute Historical Collections* 128 (1992): 286–307.

29. For "the seed of Deceit," see Daniel Leeds, *An Almanack . . . 1704* (New-York: William Bradford, [1703]); Deborah Willis, *Malevolent Nurture: Witch-hunting and Maternal Power in Early Modern England* (Ithaca: Cornell University Press, 1995).

30. For "offerings to the devil," see Johannes Megapolensis, "The Mohawk Indians," *Annals of Albany* IX: 142; for "diabolical": Denton, *Brief Description of New York*, 8–9; for "drive out the devil": Harmen Meyndertsz van den Bogaert, "A Journey into Mohawk and Oneida Country, 1634–1635," in *In Mohawk Country: Early Narratives about a Native People*, ed. Dean R. Snow, Charles T. Gehring, and William A. Starna (Syracuse, NY: Syracuse University Press, 1988), 10; for "frightfull noise": William Hyde, "The Hyde Manuscript: Captain William Hyde's Observations of the 5 Nations of Indians at New Yorke, 1698," ed. William N. Fenton, *American Scene* 6 (1965): n.p.

31. For "idols," see Van den Bogaert, *Journey*, 5; for ghosts: "Hyde Manuscript," n.p.

32. For quotations, see John Sharpe, "The Negro Plot of 1712," *New York Genealogical and Biographical Record* 21 (1890): 162–163; for Akan conjuring: Walter Rucker, "Conjure, Magic, and Power: The Influence of Afro-Atlantic Religious Practices on Slave Resistance and Rebellion," *Journal of Black Studies* 32 (2001): 84–103.

33. For scoffing as ineffectual, see Denton, *Brief Description of New York*, 8–9.

34. Lorraine Daston and Katharine Park, *Wonders and the Order of Nature, 1150–1750* (New York: Zone Books, 1998).

35. For "Sea-Lyon," see *NYG*, 17 June 1734; for kitten: *NYG*, 18 April 1737; for lion at fair: *NYG*, 29 April 1728; for Connecticut farmer: Judith Maureen Adkins, "Bodies and Boundaries: Animals in Early American Experience" (PhD diss., Yale University, 1998), 13.

36. For radish-hand story, see *NYG*, 15 May 1732.

37. Donald E. Stanford, "The Giant Bones of Claverack, New York, 1705," *New York History* 40 (1959): 47–61; Paul Semonin, *American Monster: How the Nation's First Prehistoric Creature Became a Symbol of National Identity* (New York: New York University Press, 2000), 15–40.

CHAPTER 4   *Improvement*

1. Joseph Morgan to the Lords of Trade, 5 Aug. 1714, and postscript 28 Aug. 1714, in *Documents Relating to the Colonial History of the State of New Jersey*, ed. William A. Whitehead (Newark: Daily Advertiser Printing House, 1882), IV: 190–194, 194–195.
2. For belief in decline, see comment on interpreting eclipses, Daniel Leeds, *An Almanack . . . 1696* (New-York: William Bradford, 1696); for "Green headed Time": *NYG*, 8 Nov. 1736; for chronology: Roger More, *Poor Roger, 1765: The American Almanack . . . 1765* (New-York: James Parker and John Holt, n.d.).
3. David Spadafora, *The Idea of Progress in Eighteenth-Century Britain* (New Haven, CT: Yale University Press, 1990); Larry Stewart, *The Rise of Public Science: Rhetoric, Technology, and Natural Philosophy in Newtonian Britain, 1660–1750* (Cambridge: Cambridge University Press, 1992).
4. Whitfield J. Bell, "The Reverend Mr. Joseph Morgan, An American Correspondent of the Royal Society, 1732–1739," American Philosophical Society, *Proceedings* 95 (1951): 254–261; as miller: Richard Schlatter, "Introduction," to Joseph Morgan, *The History of the Kingdom of Basaruah, and Three Unpublished Letters* (Cambridge, MA: Harvard University Press, 1946), 13.
5. For dredging: Joseph Morgan to the Royal Society, August, [1732?], Royal Society of London, "Letters and Communications from Americans, 1662–1900," microfilm, APS, reel II; for winch: Joseph Morgan to the Royal Society, Feb. 14, 1732/33, "Letters and Communications," reel II; for swift streams boat: Joseph Morgan to the Royal Society, 3 July 1734,"Letters and Communications," reel II; for locks and combination raft/canoe: [Joseph Morgan], "A Method to make shallow fresh Rivers navigable, for Skows, Flats, Barges, and such like Vessels," *Pennsylvania Gazette*, 11 and 18 May 1732.
6. For "him that Almighty God," see Morgan to the Royal Society, 5 Aug. 1714, New Jersey Historical Society, New Jersey Archives. First series. *Documents Relating to the Colonial History of the State of New Jersey*. 10 vols. (Newark, NJ: Daily Journal, 1881–1917), IV: 190–194; Ruth Bloch, *Visionary Republic: Millenial Themes in American Thought, 1756–1800* (Cambridge: Cambridge University Press, 1985), 12.
7. For "Method to Oblige," see Morgan to the Lords of Trade, 14 Feb. 1732/33, in "Letters and Communications," reel II.
8. There is a huge literature on the Royal Society, including Stewart, *Rise of Public Science*, 3–30; Steven Shapin, *A Social History of Truth: Civility and Science in Seventeenth Century England* (Chicago: University of Chicago Press, 1994).
9. For "small children," see Morgan to the Royal Society, 5 Aug. 1714, DRCHSNJ IV: 190–194; for let him know: Morgan to the Royal Society, 3 July 1734, "Letters and Communications," reel II.

10. For Morgan in Franklin's paper, see *Pennsylvania Gazette*, 11 and 18 May, 1732; for Franklin's circle: Nina Reid-Maroney, *Philadelphia's Enlightenment: Kingdom of Christ, Empire of Reason* (Westport, CT: Greenwood Press, 2001); for plough: Lewis Morris to John Morris, 25 May 1730, *The Papers of Lewis Morris*, ed. Eugene R. Sheridan (Newark: New Jersey Historical Society, 1991), I: 381–383; for quadrant: Cadwallader Colden to Peter Collinson, n.d. but with 1741 papers, NYHS *Collections*, 51 (1918): 208–211; for printing: Cadwallader Colden to Peter Collinson, 13 Nov. 1742, NYHS *Collections*, 51 (1918): 277–283; for Deane: *NYM*, 10 Feb. 1766.

11. For fire grate, see *NYG*, 1 June 1730; for desalinization: *NYWPB*, 22 April 1754; for canvas tunic: *NYM*, 8 Aug. 1763; for canals and improved navigation: *NYEP*, 23 April 1750; *NYEP*, 6 Aug. 1750; for paper mills: Hugh Gaine, *The Journals of Hugh Gaine, Printer*, ed. Paul Leicester Ford (New York: Dodd, Mead, and Co., 1902), I: 44; for Anthony Lamb: Silvio Bedini, "At the Sign of the Compass and Quadrant: The Life and Times of Anthony Lamb," American Philosophical Society, *Transactions* 74 (1984): 1–84; for Hansenclever: Irene D. Neu, "The Iron Plantations of Colonial New York," *New York History* 33 (1952): 3–24; for new immigrants: *NYWPB*, 2 Aug. 1764.

12. Elizabeth A. Fenn, *Pox Americana: The Great Smallpox Epidemic of 1775–82* (New York: Hill and Wang, 2001), 15–42.

13. Jonathan B. Tucker, *Scourge: The Once and Future Threat of Smallpox* (New York: Atlantic Monthly Press, 2001), 15.

14. Arthur Bernon Tourtellot, *Benjamin Franklin: The Shaping of Genius, The Boston Years* (Garden City, NY: Doubleday, 1977), 233–274; Tucker, *Scourge*, 16–18.

15. William Douglass to Cadwallader Colden, 1 May 1721, NYHS *Collections*, 50 (1917): 141; *NYG*, 6 May 1728.

16. *NYG*, 27 July 1730; *NYG*, 26 March 1731; *NYG*, 23 August 1731; *NYG*, 27 September 1731; the Rev. Cornelius van Schie to Messrs. Van de Wall, Beels, et al., 7 May 1732, in *ERSNY* 4: 2589–2591; James Alexander to Cadwallader Colden, 23 March 1731/32, NYHS *Collections*, 51 (1918): 59–60; *NYG*, 11 January 1731/32; *NYG*, 6 September 1731.

17. Alexander to Colden, 23 December 1731, NYHS *Collections*, 51 (1918): 39–42; *NYG*, 11 January 1731/32; Alexander to Colden, 23 March 1731/32, NYHS *Collections*, 51 (1918): 59.

18. For General Assembly meeting: *NYG*, 2 April 1739; for "keeps the Country people": Abigail Franks to Naphtali Franks, 4 December 1746, in *The Lee Max Friedman Collection of American Jewish Colonial Correspondence: Letters of the Franks Family, 1733–1748*, ed. Leo Hershkowitz and Isidore S. Meyer (Waltham, MA: American Jewish Historical Society, 1968), 134; for "warm and unchristian Debate": *NYG*, 12 December 1738; for further contentions: Alexander to Colden, 20 January 1745/46, NYHS *Collections*, 52 (1919): 190–191.

19. Samuel Johnson to William Samuel Johnson, 7 February 1763, in *Samuel Johnson, President of King's College: His Career and Writings*, ed. Herbert Schneider and Carol Schneider, 4 vols. (New York: Columbia University Press, 1929), 1: 334–335; Colden to Dr. John Bard, 5 July 1758, NYHS *Collections*, 54 (1921): 241; Donald R. Hopkins, *Princes and Peasants: Smallpox in History* (Chicago: University

of Chicago Press, 1983), 10–11, 32–33; Colden to Hugh Graham, 1716, reprinted in Saul Jarcho, "The Correspondence of Cadwallader Colden and Hugh Graham on Infectious Fevers (1716–1719)," *Bulletin of the History of Medicine* 30 (1956): 195–211; Colden to Bard, 5 July 1758, NYHS *Collections*, 54 (1921): 245.

20. *NYWPB*, 16 March 1752; *NYWPB*, 10 March 1760; *NYM*, 16 February 1761; John Nathan Hutchins, *Hutchins's Improved: Being an Almanack . . . For the Year of Our Lord, 1768* (New York: Gaine, n.d.); Joyce Goodfriend, "New York City in 1772: The Journal of Solomon Downe, Junior" *New York History* 82 (2001): 25–52; Colden to Dr. John Bard, 5 July 1758, NYHS *Collections*, 54 (1921): 236; *NYWPB*, 31 December 1759.

21. For "by carelessness," see Alexander to Colden, 20 January 1745/46, NYHS *Collections*, 52 (1919): 42; Samuel Johnson to William Samuel Johnson, 20 July 1761, *Samuel Johnson* I: 309.

22. *NYM*, 25 Feb. 1760.

23. For "a natural Fermentation," see *NYG*, 12 December 1738; *NYWPB*, 6 December 1756; Hopkins, *Princes and Peasants*, 60–61; for "Seeds of the Distemper": *NYM*, 11 February 1760; for "GOD in his mercy": *NYM, 16 February 1761.

24. For the first mathematics school, see *NYG*, 31 Aug. 1730; for example of four schools in New York City advertising in 1765 in *NYM*, 15 April, 22 April, 13 May; and 14 Oct.; for Hackensack: *NYJ*, 13 April 1769; for near Newburgh: *NYM*, 4 Oct. 1762; for example of tables: Thomas Moore, *The New-York Pocket Almanack for the Year 1768* (New-York: Hugh Gaine, n.d.); for numeracy: Patricia Cline Cohen, *A Calculating People: The Spread of Numeracy in Early America* (Chicago: University of Chicago Press, 1982); for smallpox statistics: *NYG*, 16 Oct. 1738; James Alexander to Cadwallader Colden, 20 Jan. 1745/46, in NYHS *Collections* 52 (1919): 190–191; *NYWPB*, 9 March 1751/52; *NYWPB*, 27 April 1752; *NYWPB*, 13 July 1752; *NYEP*, 6 July 1752; *NYEP*, 20 July 1752; *NYEP*, 13 Aug. 1752; *NYM*, 10 March 1760; *NYWPB*, 10 March 1760; *NYM*, 2 Feb. 1761; for numbers as facts: Mary Poovey, *A History of the Modern Fact: Problems in the Knowledge in the Sciences of Wealth and Society* (Chicago: University of Chicago Press, 1998).

25. Alexander to Colden, 23 March 1731/32, NYHS *Collections*, 51 (1918): 59–60; Alexander to Colden, 20 Jan. 1745/46, NYHS *Collections*, 52 (1919): 190–191.

26. Clinton's ban: *NYWPB*, 15 June 1747.

27. For boarders: *NYG*, 23 Oct. 1738; Samuel Johnson to William Samuel Johnson, 30 Jan. 1757, in *Samuel Johnson* I: 270–271; *NYWPB*, 27 Nov. 1752; for Muirson: *NYWPB*, 23 Dec. 1758; for Rogers: *NYJ*, 13 Oct. 1768; *NYJ*, 4 April 1771; for travel from other colonies: Benjamin Gale, "Historical Memoirs, relating to the practice of inoculation for the Small pox in the British American Provinces, particularly in New England" (1764), APS, "Letters and Communications from Americans," reel 8; Goodfriend, "New York City in 1772."

28. *NYJ*, 18 October 1770; David van Zwanenberg, "The Suttons and the Business of Inoculation," *Medical History* 22 (1978): 71–82; *NYJ*, 14 Nov. 1771; *RNYG*, 14 Oct. 1774; *NYJ*, 4 April 1771.

29. For "improving this Wilderness," see Cadwallader Colden to Gronovius, 1745, NYHS *Collections*, 52 (1919): 97–98; for admiration of Morris: Dr. Benjamin

Bullivent, "A Glance at New York in 1697: The Travel Diary of Dr. Benjamin Bullivent," ed. Wayne Andrews, NYHS *Quarterly* 40 (1956): 55–73; for the Schuylers: Lord Adam Gordon, "Journal of an Officer Who Travelled in America and the West Indies in 1764 and 1765," *Travels in the American Colonies*, ed. Newton D. Mereness (New York: The Macmillan Co., 1916), 417; for "much thrown out": Richard Smith, *A Tour of Four Great Rivers, the Hudson, Mohawk, Susquehanna, and Delaware in 1769*, ed. Francis W. Halsey (New York: Charles Scribner's Sons, 1906), 8–10; for Charles Read: Carl Raymond Woodward, *Ploughs and Politicks: Charles Read of New Jersey and His Notes on Agriculture, 1715–1774* (New Brunswick, NJ: Rutgers University Press, 1941); for madder: Samuel Bard to John Bard, 4 Sept. 1763, Samuel Bard to John Bard, 24 Nov. 1763, and John Bard to Samuel Bard, 13 Jan. 1765, transcripts of MSS letters, Malloch Rare Book Room, New York Academy of Medicine; see also Joyce E. Chaplin, *An Anxious Pursuit: Agricultural Innovation and Modernity in the Lower South, 1730–1815* (Chapel Hill: University of North Carolina Press, 1993).

30. For silk, see *NYWPB*, 8 Jan. 1767 through 2 July 1767; for hemp: Thomas More, *Poor Thomas Improved: Being More's Country Almanack for the Year of Christian Account, 1762* (New-York: William Weyman, n.d.); for example from *Gentleman's Magazine*: *NYWPB*, 3 Sept. 1753; for South Carolina: *NYJ*, 11 April 1771; for Ireland: *NYJ*, 20 July 1769; for *Young Man's Companion*: *NYWPB*, 20 May 1745.

31. For elm bark, irregular Indian orchard, see Smith, *Tour of Four Great Rivers*, 11, 55, 66; for Dutch communities: David Steven Cohen, *The Dutch-American Farm* (New York: New York University Press, 1992), 65–90; see also Thomas S. Wermuth, *Rip Van Winkle's Neighbors: The Transformation of Rural Society in the Hudson River Valley, 1720–1850* (Albany: State University of New York Press, 2001); for Hudson Valley: Cadwallader Colden to Peter Templeman, 6 Feb. 1761, NYHS *Collections*, 9 (1876): 60–61; for burning woods: *NYWPB*, 19 Dec. 1743; for drought: C. Colden, "An Account of the Earthquake felt in New-York, Novem 1755," Royal Society of London, *Philosophical Transactions*, 49 (1755): 443; for caterpillars: *RNYG*, 3 June 1773; for "alter the Virtues": Thomas More, *The American Country Almanack . . . 1750* (New-York: James Parker, n.d.).

32. For "behoves our Young Women," see Daniel Leeds, *The American Almanack . . . 1712* (New-York: William and Andrew Bradford, 1712); for silkworms: *NYWPB*, 9 Oct. 1752; for wool: *NYM*, 9 Sept. 1765; for linen: *NYM*, 11 Feb. 1765; for burnet and lucern: *NYWPB*, 12 April 1764; for hemp: *NYWPB*, 18 April 1765; for "good policy": Governor Tryon to the Earl of Dartmouth, 5 Jan. 1773, in *DRCHSNY* VIII: 342–346.

33. For town government, see Wermuth, *Van Winkle's Neighbors*, 32–34; for colony-wide government: Cathy Matson, "'Damned Scoundrels' and 'Libertisme of Trade': Freedom and Regulation in Colonial New York's Fur and Grain Trade," *William and Mary Quarterly*, 3rd series, 51 (1994): 389–418.

34. For encouraging hemp and iron production, see *NYG*, 14 and 21 Jan. 1734/35; Anon., *A Scheme (By Striking Twenty Thousand Pounds, Paper Money) To Encourage the Raising of Hemp, and the Manufacturing of Iron in the Province of New-York* (New-York: [William Bradford], 1737); for a letter commenting on

this pamphlet: *NYG*, 22 March 1736/37; for hemp lottery: *NYWPB*, 22 March 1764, and many other reports through *NYM*, 6 May 1765; for potash and pearl ash: *NYWPB*, 18 Nov. 1751; for effect of bounties: Cathy D. Matson, *Merchants and Empire: Trading in Colonial New York* (Baltimore: Johns Hopkins University Press, 1998), 252–253.

35. For associations for improvement, see Spadafora, *Idea of Progress*, 76–84; for associations elsewhere: *NYEP*, 20 Nov. 1749; *NYWPB*, 2 March 1752; *NYWPB*, 17 Sept. 1753; *NYWPB*, 19 Feb. 1761; *NYJ*, 1 Sept. 1768; *NYJ*, 24 Aug. 1769; for London society's prizes: *NYWPB*, 21 Aug. 1758; *NYM*, 30 July 1764; for William Alexander: Richard P. McCormick, "The Royal Society, the Grape, and New Jersey," New Jersey Historical Society, *Proceedings* 81 (1963): 75–83; for Long Island farmers: *NYJ*, 14 Jan. 1768.

36. For "deplorable State of our Trade": *NYWPB*, 29 Dec. 1764; for "scarce an Instance": *NYWPB*, 6 Dec. 1764; for "like Snails": *NYM*, 3 Dec. 1764; for New York society's premiums: *NYM*, 24 and 31 Dec. 1764.

37. For committee members, see *NYM*, 10 and 17 Dec. 1764; for truth and gentlemanly decorum: Steven Shapin, *A Social History of Truth: Civility and Science in Seventeenth-Century England* (Chicago: University of Chicago Press, 1994); for "every real Friend": *NYWPB*, 29 Nov. 1764; for association of premiums with "public-spirited Gentlemen": *NYM*, 12 Nov. 1764; for "Treatises abounding": *NYM*, 10 Dec. 1764.

38. For hemp pamphlet, see *NYM*, 10 June 1765, *NYWPB*, 13 June 1765; *NYWPB*, 3 April 1766 ; for "most considerable Persons": *NYWPB*, 21 Feb. 1765; for purchase of wheels: *NYM*, 25 Feb. 1765; for spinning school: *NYM*, 22 Aug. 1765; for cloth market: *NYWPB*, 10 Oct. 1765; for the market continuing at least through the beginning of 1768: *NYJ*, 14 Jan. 1768; for Schuyler and Young: *NYJ*, 14 Jan. 1768; for Thomas Young: Reverend Christopher Yonges and Pastor John Youngs, *Thomas Youngs, of Oyster Bay, and His Descendents* (Oyster Bay, NY: n.p., 1890); for detailed account of one meeting: *NYWPB*, 20 March 1766.

39. For these events, see Edmund S. Morgan and Helen M. Morgan, *The Stamp Act Crisis: Prologue to Revolution* (Chapel Hill: University of North Carolina Press, 1953; reprint, 1995), with "robbers and murderers" quote on p. 191; Richard Ketchum, *Divided Loyalties: How the American Revolution Came to New York* (New York: Henry Holt, 2002).

40. For efforts to reassure the London society: James Duane, William Smith Jr., Col. Morris, and Walter Rutherford to Society/Arts, 30 March 1765, Royal Society of Arts, "Selected Materials," APS, reel II; William Smith Jr., and Walter Rutherford to Peter Templeman, 10 Nov. 1766, R.S. of Arts, "Selected Materials," reel II; for complaints about restrictions on trade and internal taxes: Walter Rutherford on behalf of the Committee of Correspondence to Society/Arts, 29 March 1766, R.S. of Arts, "Selected Materials," reel II.

41. For "poisonous Bohea," see *NYWPB*, 11 Jan. 1768; for rye for coffee: both the *NYJ* and the *NYWPB*, 20 Nov. 1766; for "patriotic and frugal Spirit": *NYWPB*, 30 May 1765.

42. For examples of spinning news: *NYJ*, 16 June 1768; *NYJ*, 23 March, 11 May, 25 May, and 1 June 1769; for Connecticut woman: *NYJ*, 25 Oct. 1770; for "naked

breasted": John Nathan Hutchins, *An Almanack . . . 1756* (New-York: Hugh Gaine, n.d.); for "Would the Sex in general": *NYJ*, 16 Feb. 1769; T. H. Breen, "Narrative of Commercial Life: Consumption, Ideology, and Community on the Eve of the American Revolution," *William and Mary Quarterly* 50 (1993): 471–501.

43. For Colden, see Carole Shammas, "Cadwallader Colden and the Role of the King's Prerogative," NYHS, *Quarterly* 53 (1969): 103–126; for obstruction of land titles: Georgiana C. Nammack, *Fraud, Politics, and the Dispossession of the Indians: The Iroquois Land Frontier in the Colonial Period* (Norman, OK: University of Oklahoma Press, 1969), 64–69.

44. For these events, see Alan Tully, *Forming American Politics: Ideals, Interests, and Institutions in Colonial New York and Pennsylvania* (Baltimore: Johns Hopkins University Press, 1994), 166–182; for disputes over resuming imports: *NYWPB*, 30 July 1770.

45. A.B., *Proposals for Erecting and Encouraging a New Manufactory* (New-York: n.p., 1770).

46. For Schuyler, see Don R. Gerlach, *Philip Schuyler and the American Revolution in New York, 1733–1777* (Lincoln: University of Nebraska Press, 1964), 43–62; for Hurd: *NYJ*, 31 Dec. 1767; for Marine Society: *NYJ*, 2 July 1772; for a map the Marine Society sponsored: *RNYG*, 10 Feb. 1774; Gustavus D. S. Trask, *A Memoir of The Marine Society of the City of New York, Jan. 7, 1877* (n.p.: n.p., 1933); Fenn, *Pox Americana*, 65; for "the present state of the air": Edward Antill, "An Essay on the Cultivation of the Vine" APS, *Transactions* I (1769–1771): 120.

47. For skepticism, see *NYWPB*, 16 Sept. 1762; Mary Cooper, *The Diary of Mary Cooper: Life on a Long Island Farm 1768–1773*, ed. Field Horne (Oyster Bay, NY: Oyster Bay Historical Society, 1981), 15, 25; for Providence and weather: letter, *NYJ*, 4 Nov. 1773.

Chapter 5   *Refinement*

1. Jane Colden to Charles Alston, 21 May 1756, Special Collections, University of Edinburgh Library; Jane Colden, "Flora Nov. Eboraensis," Banksian MS 99, Botany Library, Natural History Museum, London; Frans Stafleu, *Linnaeus and the Linnaeans: The Spreading of Their Ideas in Systematic Botany, 1735–1789* (Utrecht: A. Oosthoek's Uitgeversmaatschappij N.V., 1971).

2. Richard L. Bushman, *The Refinement of America: Persons, Houses, Cities* (New York: Alfred A. Knopf, 1992); Lawrence E. Klein, "Gender, Conversation and the Public Sphere in Early Eighteenth-Century England," in *Textuality and Sexuality: Reading Theories and Practices*, ed. Judith Still and Michael Whorton (Manchester, UK: Manchester University Press, 1993).

3. Cary Carson, "The Consumer Revolution in Colonial British America: Why Demand?" in *Of Consuming Interests*, ed. Cary Carson, Ronald Hoffman, and Peter J. Albert (Charlottesville: University Press of Virginia, 1994); Lorinda B. R. Goodwin, *An Archaeology of Manners: The Polite World of the Merchant Elite of Colonial Massachusetts* (New York: Kluwer Academic/Plenum Publishers, 1999).

4. Nicholas Jardine and Emma Spary, "The Natures of Cultural History," in *Cultures of Natural History*, ed. N. Jardine, J. A. Secord, and E. C. Spary (Cambridge: Cambridge University Press, 1996); Neal C. Gillespie, "Natural History, Natural Theology, and Social Order: John Ray and the 'Newtonian Ideology,'" *Journal of the History of Biology* 20 (1987): 1–49; G. S. Rousseau, "Science Books and Their Readers in the Eighteenth Century," in *Books and Their Readers in Eighteenth-Century England*, ed. Isabel Rivers (New York: St. Martin's Press, 1982); Joseph M. Levine, *Dr. Woodward's Shield: History, Science, and Satire in Augustan England* (Berkeley: University of California Press, 1977).

5. James L. Reveal, *Gentle Conquest: The Botanical Discovery of North America* (Washington DC: Starwood Publishing, 1992); Amy R. W. Meyers and Margaret Beck Pritchard, eds., *Empire's Nature: Mark Catesby's New World Vision* (Chapel Hill: Omohundro Institute for Early American History and Culture and Colonial Williamsburg Foundation, University of North Carolina Press, 1998); Andrew Cunningham, "The Culture of Gardens," in *Cultures of Natural History*.

6. For colic, see #91, Asclepias, in J. Colden, "Flora Nov. Eboraensis"; for pokeweed for cancer: John Bard to Cadwallader Colden, 23 Oct. 1742, in NYHS *Collections*, 51 (1918): 274–277; Samuel Johnson to Cadwallader Colden, 25 Feb. 1745, NYHS *Collections*, 52 (1919): 104–105; Samuel Johnson to Cadwallader Colden, 26 June 1745, NYHS *Collections*, 52 (1919): 120–122; [Cadwallader Colden], "The Cure of Cancers. From an eminent Physician at New-York," *Gentleman's Magazine* (London), 13 Jan. 1751; Richard Smith, *A Tour of Four Great Rivers, the Hudson, Mohawk, Susquehanna, and Delaware in 1769*, ed. Francis W. Halsey (New York: Charles Scribner's Sons, 1906), 38.

7. For Colden's intention to write a natural history, see C. Colden to Peter Collinson, May 1742, NYHS *Collections*, 51 (1918): 257–263; for Colden's early life: Cadwallader Colden to Peter Kalm, 4 Jan. 1751, NYHS *Collections*, 53 (1920): 258–261; for his education: Saul Jarcho, "Biographical and Bibliographical Notes on Cadwallader Colden," *Bulletin of the History of Medicine*, 32 (1958): 324; for his appointment as surveyor general: Robert Hunter to Cadwallader Colden, 18 Feb. 1719/20, NYHS *Collections*, 50 (1917): 100–101; C. Colden, *Mr. Colden's Account of the Trade of New York*, in *DRCHSNY* V: 685–690; C. Colden, *The History of the Five Indian Nations Depending on the Province of New-York in America* (1727; Reprint, Ithaca, NY: Cornell University Press, 1958); C. Colden, "An Account of the Climate and Diseases of New-York in 1723," reprinted in *American Medical and Philosophical Register* I (1810–1811): 304–310; William Burnet, "Observations of the Eclipses of the first Satellite of Jupiter," Royal Society of London, *Philosophical Transactions* 33 (1724–1725): 162–164; Mary Lou Lustig, *Robert Hunter, 1666–1734: New York's Augustan Statesman* (Syracuse: Syracuse University Press, 1983); for bewilderment over unknown plants: C. Colden to Linnaeus, 9 Feb. 1748/9, NYHS *Collections*, 53 (1920): 95–99.

8. For difficulties assimilating the influx of plants, see A. G. Morton, *History of Botanical Science* (New York: Academic Press, 1981), 118–148, 195–211; for gift-giving: Susan Scott Parrish, *American Curiosity: Cultures of Natural History in the Colonial British Atlantic World* (Chapel Hill: OIEAHC by University of North Carolina Press, 2006), 167–173; Abigail Franks to Naphtali Franks, 30

Oct. 1748, in *The Lee Max Friedman Collection of American Jewish Colonial Correspondence: Letters of the Franks Family, 1733–1748*, ed. Leo Hershkowitz and Isidore S. Meyer (Waltham, MA: American Jewish Historical Society, 1968), 139–143.

9. Lisbet Koerner, *Linnaeus: Nature and Nation* (Cambridge: Harvard University Press, 1999); Stafleu, *Linnaeus and the Linnaens*, 203, 205, 211; William J. Hoffman, "The Ancestry of Rev. Gualtherus du Bois and Two Generations of His Descendents," *New York Genealogical and Biographical Record* 82 (1951): 134–139.

10. For showed the books, see C. Colden to [probably Robert Whytt], 15 Feb. 1758, NYHS *Collections*, 54 (1921): 215–217; Isaac DuBois to C. Colden, no date, Colden Scientific Manuscripts, microfilm, NYHS; for the politics of the 1720s: Patricia U. Bonomi, *A Factious People: Politics and Society in Colonial New York* (New York: Columbia University Press, 1971), 82–100; Alan Tully, *Forming American Politics: Ideals, Interests, and Institutions in Colonial New York and Pennsylvania* (Baltimore: Johns Hopkins University Press, 1994), 58–63; Jacquetta M. Haley, "Farming on the Hudson Valley Frontier: Cadwallader Colden's Farm Journal, 1727–36," *Hudson Valley Regional Review* 6 (1989): 1–34; C. Colden, Colden Scientific Manuscripts, microfilm, NYHS.

11. Colden to Peter Collinson, 13 Nov. 1742, NYHS *Collections*, 51 (1918): 277–283; Peter Collinson to C. Colden, 9 March 1743/4, NYHS *Collections*, 52 (1919): 50–52; Peter Collinson to C. Colden, 23 Aug. 1744, NYHS *Collections*, 52 (1919): 68–70; for visit: John Bartram to Peter Collinson, 5 Sept. 1742, *The Correspondence of John Bartram, 1734–1777*, ed. Edmund Berkeley and Dorothy Smith Berkeley (Gainesville: University Press of Florida, 1992), 202–203; for complaints of New Yorkers' uncooperativeness: John Bartram to Peter Collinson, 18 Oct. 1741, *Correspondence of John Bartram*, 170–172; John Bartram to John Clayton, 1 Sept. 1744, *Correspondence of John Bartram*, 244–255; John Bartram to Peter Collinson, [Fall, 1753], *Correspondence of John Bartram*, 357–364; Thomas P. Slaughter, *The Natures of John and William Bartram* (New York: Alfred A. Knopf, 1996); for the London circle: John Gascoigne, *Joseph Banks and the English Enlightenment: Useful Knowledge and Polite Culture* (Cambridge: Cambridge University Press, 1994), 76–78; Gronovius to C. Colden, 6 Aug. 1743, Gronovius to C. Colden, 3 April 1744, C. Colden to Gronovius, n.d., C. Colden to Gronovius, 29 Oct. 1745, C. Colden to Gronovius, 30 May 1746, all in "Selections from the Scientific Correspondence of Cadwallader Colden," arranged by Asa Gray, *American Journal of Science and Arts* 44 (1843): 85–133; Cadwallader Colden, "Plantae Coldenghamiae in Provincia Novaboracensi Americes sponte crescents," *Acta Societates Regiae Scientarum Upsaliensis* IV (1743): 81–136, V (1744–1750): 47–82; for the first plant in the genus *Coldenia,* a trailing plant found in Ceylon: Linnaeus, *Flora Zeylanica: sistens Plantas indicas Zeylonae insulae* (Homiae: L. Salvii, 1747), 28; for Colden informed of honor: J. Bartram to C. Colden, 6 Oct. 1746, *Correspondence of John Bartram*, 281–283; C. Colden to Linnaeus, 9 Feb. 1748/49, NYHS *Collections*, 53 (1920): 95–99; for letter writing in context of natural history: Parrish, *American Curiosity*, 106–108.

12. Peter Kalm, *Travels in North America: The English Version of 1770*, ed. and trans. Adolph B. Benson, 2 vols. (New York: Wilson Erikson, 1937); Esther Louise Larson, trans., "Peter Kalm's Short Account (1751)," *Agricultural History* 13 (1939): 33–64; for Kalm: Stafleu, *Linnaeus and the Linnaens*, 148; on Kalm's visits to Colden: C. Colden to Gronovius, 1 Oct. 1755, in "Selections from the Scientific Correspondence," 102–105; C. Colden to Linnaeus, 9 Feb. 1748/9, NYHS *Collections*, 53 (1920): 95–99; Colden's copy is now in the Malloch Rare Book Room, New York Academy of Medicine; examples of information Kalm got from New Yorkers: *Travels in North America* I: 128, I: 227, II: 437–448, II: 606; for the Rosaboms: II: 606; for their Dutch herbal by Rempert Dodoens: Alexander Hamilton, *Gentleman's Progress: The Itinerarium of Dr. Alexander Hamilton, 1744*, ed. Carl Bridenbaugh (Westport, CT: Greenwood Press, 1948), 67; for "pastime of fools": Kalm, *Travels in North America*, I: 308–309.

13. For these political events, see Stanley Nider Katz, *Newcastle's New York: Anglo-American Politics, 1732–1753* (Cambridge, MA: Harvard University Press, 1968), 165–236.

14. For Bartram visit, see John Bartram to Peter Collinson, n.d. but probably fall 1753, *Correspondence of John Bartram*, 357–364; Peter Collinson to C. Colden, 21 Nov. 1752, NYHS *Collections*, 53 (1920): 355–357; Peter Collinson to C. Colden, 10 March 1753, NYHS *Collections*, 53 (1920): 378; for bodily complaints: C. Colden to Gronovius, 1 Oct. 1755, NYHS *Collections*, 54 (1921): 29–32.

15. For Laura Bassi, see *NYG*, 31 July 1732; Gabriella Berti Logan, "The Desire to Contribute: An Eighteenth-Century Italian Woman of Science," *American Historical Review* 99 (1994): 785–812; John Mullan, "Gendered Knowledge, Gendered Minds: Women and Newtonianism, 1690–1760," in *A Question of Identity: Women, Science, and Literature*, ed. Marina Benjamin (New Brunswick, NJ: Rutgers University Press, 1993); for women's natural curiosity: C. Colden to Peter Collinson, 13 Nov. 1742, NYHS *Collections*, 51 (1918): 282; Lorraine Daston, "The Naturalized Female Intellect," *Science in Context* 5 (1992): 209–235.

16. For Jane as "Carefull": Alexander Colden (brother) to Katharine Colden (sister), 25 April 1759, NYHS *Collections*, 68 (1935): 174; for "inclination to reading": C. Colden to Gronovius, 1 Oct. 1755, NYHS *Collections*, 54 (1921): 29–32.

17. J. Colden, "Flora Nov. Eboraensis"; the Polygala is #164 in Cadwallader's flora, # 74 in Jane's; the Uvularia is #74 in Cadwallader's, #43 in Jane's; the previously unnoted plants: #153, *Hypericum virgineum*, and #292, *Fibraurea*.

18. Peter Collinson to C. Colden, 1 Sept. 1753, NYHS *Collections*, 53 (1920): 404–406; John Bartram to Peter Collinson, n.d. but probably fall 1753, *Correspondence of John Bartram*, 357–364; for publication: John Fothergill to C. Colden, 25 Oct. 1755, in *Chain of Friendship: Selected Letters of Dr. John Fothergill of London, 1735–1780*, intro and notes, Betsy C. Corner and Christopher C. Booth (Cambridge, MA: Harvard University Press, 1971), 170–173; for essay on sore throat distemper: *Medical Observations and Inquiries*, I (1757); Alexander Garden to C. Colden, 17 Dec. 1754, NYHS *Collections*, 53 (1920): 474–475; Margaret Denny, "Linnaeus and his Disciple in Carolina: Alexander Garden," *Isis* 38 (1948): 161–174.

19. For description of plant # 153 of her flora, with Edinburgh contact: Alexander Garden to John Ellis, 13 Jan. 1756, in *Selection of the Correspondence of Linnaeus*, ed. James Edward Smith (London: Longman et al., 1821), I: 362–371; "Hypericum virginium," Philosophical Society of Edinburgh, *Essays and Observations, Physical and Literary* 2 (1756): 1–7; for accuracy in identifying as not belonging to the genus Hypericum: James Britten, "Jane Colden and the Flora of New York," *Journal of Botany, British and Foreign* 33 (1895): 12–15; for conventions on naming: Stafleu, *Linnaeus and the Linnaeans*, 80–84; for Colden's new Edinburgh correspondent: C. Colden to Robert Whytt, 15 Feb. 1758, NYHS *Collections*, 54 (1921): 217; R. Emerson, "The Edinburgh Society for the Importation of Foreign Seeds and Plants," *Eighteenth-Century Life* 7 (1982): 73–95.

20. For the move and Jane's study of Latin, see C. Colden to John Fothergill, 18 Oct. 1757, NYHS *Collections*, 54 (1921): 202–204; for Colden ordering more botanical books for her: C. Colden to Peter Collinson, 31 Dec. 1757, NYHS *Collections*, 54 (1921): 211–214; for inoculation: Peter Collinson to C. Colden, 6 March 1759, NYHS *Collections*, 54 (1921): 296–298; for the marriage: *NYM*, 19 March 1759; for expectation that she would give up botany: C. Colden to Robert Whytt, Colden Scientific Manuscripts, microfilm, NYHS, reel 1; for her obituary: *NYM*, 17 March 1760; her death date is often listed erroneously as 1766; the original error: Edwin Purple, "Notes . . . of the Colden Family," *New York Genealogical and Biographical Record* 4 (1873): 170.

21. For advertisements for Spencer, see *NYWPB*, 16 and 23 Jan. 1743/44; for Brickell: *NYWPB*, 2 and 9 May 1748; for Evans: *NYWPB*, 22 and 29 July, 5 Aug. 1751; for Kinnersley: *NYWPB*, 23 March, 18 May, 15 and 22 June, and 6 July 1752; for Johnson: *NYM*, 24 Oct. and 28 Nov. 1763; *NYM*, 6 and 13 Jan. 1766; Rhode Island advice: *NYWPB*, 1 June 1752.

22. J. L. Heilbron, *Electricity in the Seventeenth and Eighteenth Centuries: A Study of Early Modern Physics* (Berkeley: University of California Press, 1979); Michael Brian Schiffer, with the assistance of Kacy L. Hollenbach and Carrie L. Bell, *Draw the Lightning Down: Benjamin Franklin and Electrical Technology in the Age of Enlightenment* (Berkeley: University of California Press, 2003); Simon Schaffer, "The Consuming Flame: Electrical Showmen and Tory Mystics in the World of Goods," in *Consumption and the World of Goods*, ed. John Brewer and Roy Porter (London: Routledge, 1993); Kevin Gumienny, "Knowledge Really Useful: Constructing and Contesting Natural Philosophy in Eighteenth-Century Philadelphia" (PhD diss., SUNY Stony Brook, 2002); for Franklin's fame in New York: *NYEP*, 3 Aug. 1752; *NYWPB*, 2 Oct. 1752; *NYM*, 6 Aug. 1753; *NYWPB*, 6 Aug. 1753; *NYWPB*, 10 Sept. 1753; *NYM*, 18 March 1754; *NYWPB*, 25 Aug. 1755; *NYWPB*, 8 July 1762; *NYM*, 12 July 1762; *NYM*, 19 Oct. 1767; *NYJ*, 21 Jan. 1768; *NYWPB*, 16 May 1768; *NYJ*, 15 June 1769; for positive reports on lightning rods that don't mention Franklin: *NYWPB*, 31 July 1766; *NYJ*, 24 Aug. 1769; *NYJ*, 8 Aug. 1771; *NYJ*, 19 Aug. 1773; for South Carolina poem: *NYWPB*, 15 Sept. 1754.

23. William Johnson, *A Course of Experiments, In that Curious and Entertaining Branch of Natural Philosophy, call'd Electricity: Accompanied with Lectures on the Nature and Properties of the Electric Fire* (New-York: H. Gaine, 1765.)

24. For calf, see *NYEP*, 16 March 1752; for monkey: *NYWPB*, 18 Feb. 1750/51; for Bonin's ads: *NYWPB*, 31 Oct. 1748, 2, 9, and 16 Jan., 13 March, and 10 April 1749; to Long Island: *NYWPB*, 6 Feb. 1748/49; lowers price: *NYWPB*, 12 Dec. 1748; promises to pay creditors: *NYWPB*, 28 May 1750; see Geoffrey V. Sutton, *Science for a Polite Society: Gender, Culture, and the Demonstration of Enlightenment* (Boulder: Westview Press, 1995).

25. For porcupine, see *NYWPB*, 9 Dec. 1751; for alligator: *NYM*, 1 July 1754; for snake: *NYWPB*, 22 Nov. 1756; for buffalo: *NYM*, 19 Dec. 1757; for "tiger": *NYWPB*, 4 June 1761; for elk: *NYWPB*, 11 Aug. 1763, and *Rivington's NYG*, 2 Sept. 1773.

26. For "Preter-naturals," see *NYEP*, 13 Jan. 1745/46; for "soft as a roasted apple": *NYJ*, 8 Aug. 1771; for epistemological decorum: Lorraine Daston, "Baconian Facts, Academic Civility, and the Prehistory of Objectivity," *Annals of Scholarship* 8 (1991): 337–363.

27. For women combusting, see *NYWPB*, 7 Jan. 1771 and *NYWPB*, 25 Feb. 1771; for man walking across water: *NYEP*, 30 June 1746; for knife in cow: *NYJ*, 30 Dec. 1773; for merman sighting: *NYWPB*, 11 March 1762; *NYM*, 15 March 1762; for Whitbourne's comments: *NYM*, 22 March 1762.

28. For similarly slow increase in doctors with medical degrees, see Eric H. Christianson, "The Medical Practitioners of Massachusetts, 1630–1800: Patterns of Change and Continuity," in *Medicine in Colonial Massachusetts, 1620–1820* (Boston: Colonial Society of Massachusetts, 1980); Nicholas Jewson, "Medical Knowledge and the Patronage System in Eighteenth-Century England," *Sociology* 8 (1974): 369–385.

29. For "stiled himself *Medicinae* Doctor," see *NYEP*, 18 Nov. 1751; for Young: Henry H. Edes, "Memoir of Dr. Thomas Young, 1731–1777," Colonial Society of Massachusetts, *Transactions* 11 (1910): 2–54; Pauline Maier, "Reason and Revolution: The Radicalism of Dr. Thomas Young," *American Quarterly* 28 (1976): 229–249; for Clinton: Charles Clinton (father) to Captain James Clinton (brother), 21 Aug. 1761, MSS letter, NYHS; Charles Clinton (father) to Charles Clinton (son), 2 July 1762, MSS letter, NYHS; for examples of doctors' ads seeking apprentices: *NYWPB*, 3 Feb. 1755; *RNYG*, 25 Aug. 1774; for shoemaker-turned-doctor: Hamilton, *Gentleman's Progress*, 91; for "every man who drives the curing trade": Johann David Schoepf, *Travels in the Confederation, 1783–1784*, ed. and trans. Alfred J. Morrison, 2 vols. (Philadelphia: William J. Campbell, 1911), I: 76.

30. Thomas More, *Poor Thomas Improved: Being More's Country Almanack . . . 1768* (New-York: William Weyman, n.d.); for example of ad for *English Physician*: *NYWPB*, 27 Aug. 1750; for *The Lady's Dispensary*: *NYM*, 25 Feb. 1754; for *Young Man's Companion* with *The Poor Planter's Physician*: *NYM*, 3 March 1760; for *Medicina Britannica*: *NYM*, 14 June 1753; for medicine chest ad: *NYM*, 4 Nov. 1754; P. S. Brown, "Medicine Advertised in Eighteenth-Century Bath Newspapers," *Medical History* 20 (1976): 152–168; Colin Jones, "The Great Chain of Buying: Medical Advertisement, the Bourgeois Public Sphere, and the Origins of the French Revolution," *American Historical Review* 101 (1996): 13–40.

31. For Murfoy, see *NYEP*, 24 Aug. 1747 and repeated fifteen times; for Levine: *NYM*, 3 May 1762; for Clark: *NYWPB*, 8 Jan. 1750; for Howard: *NYG*, 16 March 1729/30; for Richards: *NYM*, 25 May 1767; for Elkin: *NYM*, 22 June 1761; for Richardson: *NYWPB*, 3 Nov. 1763; example of ad with testimonials attached: *RNYG*, 19 Aug. 1773; ad with promise of "No Cure, No Pay": *NYWPB*, 9 Jan. 1753.

32. Report from the *Philosophical Transactions*: *NYG*, 28 May 1739; for Westchester farmer's success with salad oil: *NYG*, 13 Aug. 1739; for hog lard: Anonymous, *The Husband-man's Guide, in Four Parts*, 2nd ed. (New-York: William Bradford and Andrew Bradford, 1712), 43; for Evans's account: Colden Scientific Manuscript, microfilm, NYHS; printed, with quote: *Colden's Observations on the bite of a rattle snake*, NYHS *Collections*, 52 (1919): 66–68.

33. For the conflict between Schaw and his rivals, see *NYWPB*, 28 Oct., 4, 11, 18, and 25 Nov. 1751; *NYEP*, 18 and 25 Nov. 1751; for Schaw as practicing in New Jersey: *NYWPB*, 3 Aug. 1752.

34. For Hamilton's encounter, see Hamilton, *Gentleman's Progress*, 179–180; for Jay: Samuel Bard to John Bard, 1773, transcript of MSS letter, Malloch Rare Book Room, New York Academy of Medicine; for ferry accident: *New York Journal*, 29 Aug. 1771; for Yeldell's ad: *RNYG*, 22 June 1774.

35. Andrew Wear, "Medical Practice in Late Seventeenth and Early Eighteenth-Century England: Continuity and Union," in *Health and Healing in Early Modern England: Studies in Social and Intellectual History*, ed. Andrew Wear (Aldershot, UK: Ashgate/Variorum, 1998); Theodore M. Brown, "Medicine in the Shadow of the *Principia*," *Journal of the History of Ideas* 48 (1987): 629–648; Don G. Bates, "Thomas Willis and the Fevers Literature of the Seventeenth Century," in *Theories of Fever from Antiquity to the Enlightenment*, ed. W. F. Bynum and V. Nutton (London: Wellcome Institute for the History of Medicine, 1981); Roy Porter, *The Greatest Benefit to Mankind: A Medical History of Humanity* (New York: W. W. Norton, 1998), 201–303; for "experimental Turn": *NYWPB*, 5 Feb. 1767; for "prospect of farther discoveries": Peter Middleton, *A Medical Discourse, or A historical Inquiry into the Ancient and Present State of Medicine* (New-York: Hugh Gaine, 1769), 63.

36. For reminder to cultivate a gentleman's polish, see John Bard to Samuel Bard, 9 April 1763; for lectures on rhetoric and belles lettres: Samuel Bard to John Bard, 4 Sept. 1763; for advice on garden design: Samuel Bard to John Bard, Sept. 1763; for advice on reading: Samuel Bard to brother, 2 June 1764, all transcripts of MSS letters, Bard Collection, Malloch Rare Book Room, New York Academy of Medicine.

37. For calls for licensing physicians, see *NYEP*, 3 April 1749; *NYWPB*, 29 Oct. 1750; *Independent Reflector*, 5 Feb. and 10 May 1753; *NYM*, 30 April 1759; for law regulating midwives: *NYWJ*, 26 June 1738; on rare occasion, a midwife advertised that she had been duly examined: ad of Mrs. Fisher, *NYWPB*, 9 Jan. 1769; ad of M. Hartley, *NYJ*, 30 May 1771; for medical licensure: John Duffy, *A History of Public Health in New York City, 1625–1866* (New York: Russell Sage Foundation, 1968), 65–66; ad for Clossy's anatomy course, which he repeated every autumn subsequently: *NYWPB*, 17 Nov. 1763; for Clossy: Samuel Clossy, *The Existing Works with a Biographical Sketch by Morris Saffron* (New York:

Hafner Publishing, 1967); for the Bards' hope to imitate Philadelphia: John Bard to Samuel Bard, 9 April 1763, Bard Collection, New York Academy of Medicine; for medical school: *NYWPB*, 24 Sept. 1767; David C. Humphrey, *From King's College to Columbia, 1746–1800* (New York: Columbia University Press, 1976), 233–260; for "attentive old Woman": Peter Middleton, *A Medical Discourse, or an Historical Inquiry into the Ancient and Present State of Medicine* (New-York: Hugh Gaine, 1769).

38. For modest impact, see Humphrey, *From King's College*, 239–240, 251; for "dashing ones brains out": John Bard to Samuel Bard, 13 Jan. 1765, Bard Collection, New York Academy of Medicine; Steven Shapin, "'A Scholar and a Gentleman': The Problematic Identity of the Scientific Practitioner in Early Modern England," *History of Science* 29 (1991): 279–327.

39. For "good Pennysworths," see John Watts to Colonel Barré, 15 Nov. 1763, NYHS *Collections*, 61 (1928): 197–199; for satire on corns: *NYWJ*, 7 Sept. 1741; for ditty in newspaper: *NYWJ*, 3 Feb. 1748/49.

40. For complaint about tar water: *NYWPB*, 15 July 1745.

41. Ghislaine Lawrence, "Surgery (Traditional)" in *Companion Encyclopedia of the History of Medicine*, ed. W. F. Bynum and Roy Porter, 2 vols. (London: Routledge, 1993), II: 977, 979; for reports of Jones's successful lithotomies: *NYWPB*, 2 July 1767; *NYM*, 6 July 1767; *NYJ*, 22 Oct. 1767; *NYJ*, 11 May 1769; *NYWPB*, 21 Jan. 1771; for Jones: James Thacher, *American Medical Biography* (Boston: Richardson and Lord, 1828), 324–330; the most renowned oculist was an itinerant, Dr. Graham: first ad in New York, *NYWPB*, 20 August 1770; in New Jersey: *NYJ*, 30 June 1771; his return to New York with reports of his cures in Virginia: *RNYG*, 15 July 1773; examples of cures he performed in New York: *NYJ*, 29 July 1773.

42. Ad to "purge the Humours": *NYM*, 8 Jan. 1759; for "corrector of Juices": *NYM*, 24 Sept. 1759; for example of persistence of older language: W. Wing, *Wing Improv'd: An Almanack, For the Year of our Lord Christ, 1764* (New-York: Samuel Brown, 1764); for examples of phlegm in newer sense: *NYM*, 3 July 1758; *NYWPB*, 22 Jan. 1770; for blood being of central concern: directions for a drink to "rectify" the blood, Dr. George Muirson to Henry Lloyd, 14 March 1749/50, NYHS *Collections*, 59 (1926): 443–444; *NYM*, 1 Aug. 1763; for example of being bled, blistered, and purged: Peter Wraxall to Sir Charles Hardy, 8 Oct. 1755, in Sir William Johnson, *The Papers of Sir William Johnson*, ed. James Sullivan and Milton W. Hamilton, 13 vols. (Albany: The University of the State of New York, 1921–1962), II: 153–155.

43. For "nervous Texture of the Female Sex": *NYWPB*, 23 Jan. 1769; for "loose fibers": *NYWPB*, 19 Oct. 1772; for skin as porous: John Lloyd to Henry Lloyd, 11 March 1750/51, NYHS *Collections*, 59 (1926): 471–472; *NYWPB*, 24 Dec. 1760; *NYM*, 5 Oct. 1767; for "very Medicine that is of Service one Day": *NYJ*, 9 April, 1767.

44. For "Regulation of our Passions," see *NYWJ*, 8 Dec. 1735; for warnings against excessive alcohol: *NYWJ*, 27 Aug. 1739; *NYEP*, 11 Sept. 1749; *NYEP*, 10 June 1751; for death from "a surfiet taken by over Exercise": Henry Lloyd (son) to Henry Lloyd, 20 Jan. 1755, NYHS *Collections*, 60 (1927): 527; for deaths in hot

weather after drinking cold water: *NYWJ*, 15 July 1734; *NYEP*, 4 Aug. 1746; *NYWPB*, 13 July 1752; *NYM*, 22 June 1767; *NYWPB*, 12 Aug. 1771; *NYJ*, 15 Aug. 1771; *NYJ*, 24 Aug. 1771.

45. Poem: *NYJ*, 13 Dec. 1770; for Clossy's complaint: quoted in introduction by Morris S. Saffron to Clossy, *The Existing Works*, xxx–xxxi; for Sarah Parker story: *NYM*, 2 March 1767.

46. For resuscitation advice or descriptions: *NYWJ*, 14 April 1740; *NYWPB*, 5 Feb. 1753; *NYM*, 8 Aug. 1763; *NYWPB*, 9 Aug. 1764; for Dr. Clossy's advice: *NYJ*, 5 Aug. 1773; *RNYG*, 23 Nov. 1773; *RNYG*, 25 Aug. 1774; stories with indistinct line between life and death: *NYWPB*, 19 Feb. 1753; *NYM*, 25 May 1767; *NYWPB*, 4 July 1768; for Bergan County case: *NYJ*, 1 Oct. 1767; for other cases: *NYWPB*, 11 April 1768; J. N. Hutchins, *Hutchin's Improved: Being an Almanack . . . 1771* (New-York: Hugh Gaine, n.d.); David C. Humphrey, "Dissection and Discrimination: The Social Origins of Cadavers in America, 1760–1915," *Bulletin of the New York Academy of Medicine* 49 (1973): 819–827.

47. For Livingston, see Milton M. Klein, "Introduction," in *The Independent Reflector, or, Weekly Essays on Sundry Subjects More particularly adapted to the Province of New-York*, ed. Milton M. Klein (Cambridge, MA: Harvard University Press, 1963); [William Livingston], "Philosophic Solitude Or The Choice of a Rural Life. A Poem" (New-York: James Parker, 1747); a complaint about cockfighting: *NYJ*, 18 Feb. 1773; Serena R. Zabin, "Places of Exchange: New York City, 1700–1763" (PhD diss., Rutgers University, 2000).

48. Steven C. Bullock, "A Mumper Among the Gentle: Tom Bell, Colonial Confidence Man," *William & Mary Quarterly* 55 (1998): 231–258; for credit among New Yorkers: Cathy D. Matson, *Merchants and Empire: Trading in Colonial New York* (Baltimore: Johns Hopkins University Press, 1998), 152–155.

49. For gentleman's club, see Samuel Bard to John Bard, 1773, Bard Collection, Malloch Rare Book Room, New York Academy of Medicine; for Johnson's mansion: Rev. Gideon Hawley, "Narrative of his Journey to the Ohohoghwage, July 1753," in *ERSNY* V: 3399–3405; for his garden: William Darlington to William Johnson, 30 Oct. 1762, *William Johnson Papers* III: 918–919; for his telescope: Goldsbrow Banyar to William Johnson, 10 June 1755, *William Johnson Papers* I: 576; for deplored low education of local doctors: William Johnson to Thomas Barton, 21 Nov. 1765, *William Johnson Papers* IV: 874–897; see also Wanda Burch, "Sir William Johnson's Cabinet of Curiosities," *New York History* 71 (1990): 261–282; for Young's nursery: Ulysses Prentiss Hedrick, *A History of Agriculture in the State of New York* (Albany: J.B. Lyon for the New York Agricultural Society, 1933), 72; advertisement for Cook's book: *NYJ*, 22 Dec. 1774.

Chapter 6 *Reason*

1. Robert Hunter to the Lords of Trade, 23 June 1712, in *DRCHSNY* V: 339–343; Robert Hunter to the Lords of Trade, 12 Nov. 1715, in *DRCHSNY* V: 457–463; the pestilence/census Bible story: 2 Samuel 24.

2. For theological concerns as central, see Andrew Cunningham, "How the *Principia* Got Its Name: Or, Taking Natural Philosophy Seriously," *History of Science* 29

(1991): 377–392; Dorinda Outram, *The Enlightenment* (Cambridge: Cambridge University Press, 1995), 31–62.

3. Samuel Johnson's essays appeared in the *NYG*, 28 Jan., 4, 11, 18, and 25 Feb., 4, 11, 17, 24, and 31 March, 28 April, 5, 12, and 19 May, and 1 Sept. 1729; while they are unsigned or signed "T.P.N.," the eighth essay is among his papers: Samuel Johnson, *Samuel Johnson, President of King's College: His Career and Writings*, ed. Herbert Schneider and Carol Schneider, 4 vols. (New York: Columbia University Press, 1929), 2: 254–260; quoted material in this paragraph is from essays 2 and 3, *NYG*, 4 and 11 Feb. 1728/29; for "timorous Imagination": *NYWPB*, 12 Aug. 1751; for mockery of conjurers: W. Jones, *The American Almanack . . . 1750* (New-York: Henry De Foreest, n.d.); for mockery of fortune-tellers: *NYWPB*, 18 March 1751; *NYEP*, 8 June 1752.

4. [Samuel Johnson], Essay 14, *NYG*, 19 May 1729; for kite story: *NYG*, 22 Nov. 1731; for earthquake with saint story: *NYM*, 17 March 1760; for similar stories mocking Catholic "superstition": *NYWPB*, 11 Dec. 1749; *NYJ*, 16 Jan. 1772; "Merry Andrew, Stargazer," *Merry Andrew's New Almanack . . . 1774* (New-York: John Anderson, n.d.).

5. For mocking "enthusiasts," see *NYWPB*, 13 Aug. 1750; for "imbued with the Doctrines": William Livingston, "Evils of a Sectarian College supported by Public Funds," 23 April 1753, in *ERSNY* V: 3339–3341; for "Miseries of Priest-Craft": William Livingston, "Appeal to the Inhabitants of New York against a Sectarian College," in *ERSNY*: V, 3366–3369; for the King's College controversy: David C. Humphrey, *From King's College to Columbia, 1746–1800* (New York: Columbia University Press, 1976), 18–66.

6. For Britain, see Margaret Jacob, *The Cultural Meaning of the Scientific Revolution* (Philadelphia: Temple University Press, 1988), 105–131; Neal C. Gillespie, "Natural History, Natural Theology, and Social Order: John Ray and the 'Newtonian Ideology,'" *Journal of the History of Biology* 20 (1987): 1–50; for Kinnersley: Nina Reid-Maroney, *Philadelphia's Enlightenment, 1740–1800: Kingdom of Christ, Empire of Reason* (Westport, CT: Greenwood Press, 2001), 52–60; Henry Melchior Muhlenberg, *The Journals of Henry Melchior Muhlenberg*, trans. Theodore G. Tappert and John W. Doberstein, 3 vols. (Philadelphia: The Muhlenberg Press, 1942), II: 159.

7. For Neoplatonism and Hermeticism, see John L. Brooke, *The Refiner's Fire: The Making of Mormon Cosmology, 1664–1844* (Cambridge: Cambridge University Press, 1994), 8–25; for recipes relying on "sympathy": Thomas More, *The American Country Almanack . . . 1752* (New-York: James Parker, n.d.); Roger More, *Poor Roger. The American Country Almanack . . . 1761* (New-York: James Parker & Co., n.d.); for anodyne necklace ad: *NYWPB*, 22 June 1747; for amber beads cure: John Aspinwall to Joseph Lloyd II, 30 Oct. 1772, NYHS *Collections*, 60 (1927): 733; for Hartford pregnancy story: *NYWPB*, 10 Sept. 1770, and *NYJ*, 13 Sept. 1770; *NYWPB*, 9 Aug. 1764; for Johnson's list: *NYG*, 4 Feb. 1728/29.

8. For Johnson's skepticism, see Essay 2, *NYG*, 4 Feb. 1728/29; for similar expressions: *NYG*, 25 Sept. 1732; Roger Sherman, *An Almanack . . . 1752* (New-York: William De Foreest, n.d.); *NYWPB*, 8 April 1754; Mr. Prion, "The Vanity

of attempting Supernatural Knowledge," *New American Magazine*, Jan. 1758 (Woodbridge, NJ: James Parker); Roger More, *Poor Roger. The American Country Almanack . . . 1768* (New-York: James Parker, n.d.).

9. Titan Leeds, *The American Almanack . . . 1732* (New-York: William Bradford, n.d.); for Johnson: *NYG*, 4 Feb. 1728/29; *NYG*, 18 Sept. 1732; Jesse Parsons, *The American Ephemeris, or, an Almanack . . . 1757* (New-York: J. Parker and W. Weyman, n.d.); Roger More, *Poor Roger. The American Country Almanack . . . 1773* (New-York: Samuel Inslee and Anthony Car, n.d.); for Wilkins: Steven Shapin, *A Social History of Truth: Civility and Science in Seventeenth-Century England* (Chicago: University of Chicago Press, 1994), 198–199; see also Sara Schechner Genuth, "Devils' Hells and Astronomers' Heavens: Religion, Method and Popular Culture in Speculations about Life on Comets," in *The Invention of Physical Science*, ed. Mary Jo Nye, Joan L. Richards, and Roger Stuewer (Dordrecht: Kluwer Academic, 1992).

10. For "INIQUITIES," see *NYWPB*, 21 May 1750; for poems: *NYM*, 17 May 1756; for Colden's account: C. Colden, "An Account of the Earthquake felt in New York, Novem.18, 1755, in a letter from Cadwallader Colden, Esq.; to Mr. Peter Collinson, FRS," Royal Society of London, *Philosophical Transactions* 49 (1755): 443; for Mohawk Valley account: John Ogilvie, "Diary, 1750–1759," ed. Milton W. Hamilton, *Bulletin of the Fort Ticonderoga Museum* 10 (1961): 366; for other examples in which earthquakes were interpreted providentially: *NYG*, 19 Dec. 1726; *NYEP*, 2 July 1750; *NYWPB*, 23 Feb. 1756; *NYM*, 29 Nov. 1756; for plain reports: *NYWPB*, 20 March 1761; *NYWPB*, 11 Sept. 1766; *NYJ*, 12 July 1770; for earthquakes reported along with accompanying phenomena or naturalistic explanations: *NYEP*, 30 April 1750; *NYEP*, 24 Feb. 1752; *NYWPB*, 29 April 1754; for essay on earthquakes: *New American Magazine*, June, July, and Aug. 1758; for continued providential interpretations in Britain, particularly in the 1750s: David Spadafora, *The Idea of Progress in Eighteenth-Century Britain* (New Haven: Yale University Press, 1990), 109–119; for aurora borealis: Daniel Claus to William Johnson, 26 Feb. 1761, in *The Papers of Sir William Johnson*, ed. James Sullivan and Milton W. Hamilton, 13 vols. (Albany: University of the State of New York, 1921–1962), III: 348–349.

11. For examples of the foolish accusation of old women, see *NYG*, 18 July 1726; *NYG*, 26 April 1731; *NYWJ*, 12 December 1737; for a lady who intervened: *NYG*, 3 April 1738; *NYWJ*, 5 June 1738; *NYEP*, 15 July 1751; *NYWPB*, 12 Aug. 1751; *NYEP*, 13 July 1752; for "country People": *NYWPB*, 1 Jan. 1753; for studies of two such cases: Phyllis J. Guskin, "The Context of Witchcraft: The Case of Jane Wenham (1712)," *Eighteenth-Century Studies* 15 (1981): 48–71; and W. B. Carnochan, "Witch-Hunting and Belief in 1751: The Case of Thomas Colley and Ruth Osborne," *Journal of Social History* 4 (1971): 389–403; for sow and "genteel perspiration": *NYM*, 9 March 1761; for the general decline in witchcraft persecutions: Ian Bostridge, *Witchcraft and Its Transformation, c. 1650–c. 1750* (Oxford: Clarendon Press, 1997); Willem Frijoff, "The Emancipation of the Dutch Elites from the Magic Universe," in *The World of William and Mary: Anglo-Dutch Perspectives on the Revolution of 1688–89*, ed. Dale Hoak and Mordechai Feingold (Stanford: Stanford University Press, 1996).

12. For "Cart Loade of old Dutch people," see John Lyne to William Johnson, 25 Feb. 1745/46, in *Papers of Sir William Johnson* I: 44–45; for Smith's experiences: William Hooker Smith, "Remarkable acurances," manuscript with typescript in "Westchester" box, Special Collections, NYPL.

13. Two excellent studies that emphasize the absence of modern disciplinary boundaries: Rhoda Rappaport, *When Geologists Were Historians, 1665–1750* (Ithaca: Cornell University Press, 1997); Patricia Fara, *Sympathetic Attractions: Magnetic Practices, Beliefs, and Symbolism in Eighteenth-Century England* (Princeton: Princeton University Press, 1996).

14. For the centrality of theology to natural philosophy, see Larry Stewart, *The Rise of Public Science: Rhetoric, Technology, and Natural Philosophy in Newtonian Britain, 1660–1750* (Cambridge: Cambridge University Press, 1992), 31–59.

15. For "wonderful Discoveries," see *NYG*, 17 March 1728/29; for statues: *NYG*, 16 Aug. 1731; *NYG*, 4 March 1739/40; for "Immortal Newton!": [William Livingston], "Philosophic Solitude: Or, The Choice of a Rural Life. A Poem" (New-York: James Parker, 1747); for Zenger's plea: *NYWJ*, 19 June 1749; for ignorant Albany doctor: Alexander Hamilton, *Gentleman's Progress: The Itinerarium of Dr. Alexander Hamilton, 1744*, ed. Carl Bridenbaugh (Westport, CT: Greenwood Press, 1948), 85–86.

16. Joseph Morgan, "Some farther Improvemt on ye Astronomical Philosophy of ye Astronomical Philosophy of Sir Isaac Newton & others," enclosed with "The Originals of all Nations," undated but ca. 1737, Royal Society of London, "Letters and Communications from Americans, 1662–1900," APS, reel III; for encounter with Morgan: Hamilton, *Gentleman's Progress*, 36.

17. Colden's student notes showed he had studied Newton: Brooke Hindle, "Cadwallader Colden's Extension of the Newtonian Principles," *William and Mary Quarterly*, 3rd series, 13 (1956): 462; C. Colden to Peter Collinson, 20 June 1745, NYHS *Collections*, 52 (1919): 117–118; Cadwallader Colden, *An Explication of the First Causes of Action in Matter and the Causes of Gravitation* (New York: James Parker, 1745).

18. For resistance as an active power, see Colden, *Explication of First Causes*, 4; for discussion of his ideas of resistance: "Scientific Note," NYHS *Collections*, 52 (1919): 157; for discussion of the ideas of Descartes, Malabranche, Newton, and Boerhaave on light: John Rutherford to C. Colden, 2 March, 1742/43, NYHS *Collections*, 52 (191): 6–10; J. Rutherford to C. Colden, 19 April 1743, NYHS *Collections*, 52 (1919): 17–21; Johnson sent him George Berkeley's *New Theory of Vision*: S. Johnson to C. Colden, 21 Nov. 1743, NYSH, Col., 52 (1919): 39; for aether: J. L. Heilbron, *Elements of Early Modern Physics* (Berkeley: University of California Press, 1982): 55–60.

19. For "divine Gouvenor," see handwritten chapter 8 for second edition of *The Principles of Actions in Matter*, microfilm "Colden Material," EUL Dc1.25–26, Edinburgh University Library; for the difficulties applying Newtonianism to politics: Richard Striner, "Political Newtonianism: The Cosmic Model of Politics in Europe and America," *William and Mary Quarterly*, 3rd series, 52 (1995): 583–608; for Colden's belief that his work would illuminate many phenomena: *The Principles of Action in Matter, the Gravitation of Bodies, and the Motion of the Planets,*

explained from these Principles (London: R. Dodsley, 1751), chap. 6; for the application to the circulation of the blood: C. Colden to Robert Whytt, 2 Feb. 1765, NYHS *Collections*, 9 (1876): 464–465; for the axiom that divine wisdom ensured the fundamental simplicity of nature: J. H. Brooke, "Why did the English Mix Their Science and Their Religion?" in *Conference on Science and Imagination in British Culture during the Eighteenth Century*, ed. Sergio Rossi (Milan, Edizioni Unicopl, 1987).

20. For promise of a simpler mathematics, see C. Colden, *Explication of the First Causes*, 42; for only arithmetic and trigonometry required: C. Colden, *Principles of Action in Matter*, preface; for expectation of use to astronomy, navigation, and geography: C. Colden to P. Collinson, 20 June 1745, NYHS *Collections*, 52 (1919): 118–119; Colden also apparently wrote a short essay on the calculus that circulated among his friends: John Rutherford to C. Colden, 2 April 1743, NYHS *Collections*, 52 (1919): 6–9; but apparently Colden did not completely understand it: Benjamin Franklin to C. Colden, 25 Oct. 1744, NYHS *Collections*, 52 (1919): 77–78; for Newton's mathematics and astronomy: Bruce Chandler, "Longitude in the Context of Mathematics," *The Quest for Longitude: The Proceedings of the Longitude Symposium, Harvard University, Nov. 4–6, 1993*, ed. William J. H. Andrewes (Cambridge, MA: Collection of Historical Scientific Instruments, Harvard University, 1996); for the difficulties in mastering the calculus using Newton's notation and the consequent lag in Britain in the skilled use of the calculus: Dirk J. Struik, *A Concise History of Mathematics*, 4th revised edition (New York: Dover Publications, 1987), 130; for "Conquest of Kingdoms": James Alexander to C. Colden, 29 March 1748, NYHS *Collections*, 53 (1920): 30.

21. J. Alexander to C. Colden, 30 Jan. 1745/46, NYHS *Collections*, 52 (1919): 195–196; John Rutherford to C. Colden, 2 March 1742/43, NYHS *Collections*, 52 (1919): 6–9; for admiration mixed with reservations: S. Johnson to C. Colden, 10 July 1745, NYHS *Collections*, 52 (1919): 127–128; B. Franklin to C. Colden, 15 Aug. 1745, NYHS *Collections*, 52 (1919): 139–140; sent to "Greatest People": P. Collinson to C. Colden, 27 March 1746/47, NYHS *Collections*, 52 (1919): 368; for problems understanding it: B. Franklin to C. Colden, 16 Oct. 1746, NYHS *Collections*, 52 (1919): 273–276; for Clapp: S. Johnson to C. Colden, 12 Jan. 1746/47, NYHS *Collections*, 52 (1919): 330–332; for Franklin reporting he had heard from two sources that Colden's treatise got a "great Reception in England": B. Franklin to C. Colden, 6 Aug. 1747, NYHS *Collections*, 52 (1919): 414–415; for German translation: P. Collinson to C. Colden, 20 Aug. 1748, NYHS *Collections*, 53 (1920): 67; for English astronomer: John Betts to C. Colden, 25 April 1750, NYHS *Collections*, 53 (1920): 204–207.

22. For "uncommonness and newness," see C. Colden to S. Johnson, 19 Nov. 1746, NYHS *Collections*, 52 (1919): 282–283; ordered Colin McLaurin's *Newtonian Philosophia*: C. Colden to Thomas Osborne, 1748, NYHS *Collections*, 56 (1923): 344–345; for McLaurin: Robert E. Schofield, *Mechanism and Materialism: British Natural Philosophy in an Age of Reason* (Princeton: Princeton University Press, 1970), 102–109; ordered George Berkeley's *Siris* and Benjamin Martin's *Philosophia Britannica*: C. Colden to P. Collinson, 7 July 1749, NYHS

Collections, 53 (1920): 114–116; these two books dispatched: P. Collinson to C. Colden, 17 Aug. 1749, NYHS *Collections*, 53 (1920): 131; the problem Colden couldn't solve was the "three-body" problem: C. Colden to J. Betts, 25 April 1750, NYHS *Collections*, 53 (1920): 204–207; Geoffrey Sutton, *Science for a Polite Society: Gender, Culture, and the Demonstration of Enlightenment* (Boulder, CO: Westview, 1995), 266–273.

23. For an example of simile, see Colden, *Explication of First Causes*, 35; the source for his mathematics was Wallis's *Arithmetica Infinitorum*: manuscript revision of chapter two, *Principles of Action in Matter*; Colden repeatedly asserted that using others' observations was the equivalent of "experience": C. Colden to J. Betts, 25 April 1750, NYHS *Collections*, 53 (1920): 204–207; C. Colden to B. Franklin, 2 April 1754, NYHS *Collections*, 53 (1920): 437.

24. For Euler's comments conveyed, see P. Collinson to C. Colden, 21 Nov. 1752, NYHS *Collections*, 53 (1920): 355–357; for another critique: C. Colden to B. Franklin, 16 March 1752, NYHS *Collections*, 53 (1920): 314–317; Colden collated several criticisms of his treatise: two documents, both labeled "Criticisms of Principles of Action," Colden Scientific Manuscripts, microfilm, NYHS; for Ovid quote: C. Colden to B. Franklin, 20 May 1752, NYHS *Collections*, 53 (1920): 325–328; Ovid, "Epitaph for Phaethon," *The Metamorphoses*, Book 2, line 328, trans. Mary M. Innes (Hammonsworth, UK: Penguin Books, 1955); my thanks to Serena R. Zabin for identifying the quotation.

25. For the little understanding of mathematics among British readers of Newton: Heilbron, *Elements of Early Modern Physics*, 9; for "wrote in an analytic Method": *London Magazine* 21 (1752): 560–562; reviewed somewhat more critically: *The Monthly Review* 7 (1952): 459–467; "cleared up by a Countryman": "Criticisms of Principles of Action," Colden Scientific Ms.

26. For exchanges on the distinction between intentionless matter and intelligent matter, see C. Colden to S. Johnson, 12 April 1746, NYHS *Collections*, 52 (1919): 202–205; S. Johnson to C. Colden, 22 April 1746, NYHS *Collections*, 52 (1919): 205–208; C. Colden to S. Johnson, 19 Nov. 1746, NYHS *Collections*, 52 (1919): 281–283; S. Johnson to C. Colden, 6 June 1747, NYHS *Collections*, 52 (1919): 398–400; for "such a notion lead": "J. Bevis's Remarks on Cadwallader Colden's *Principles*," 10 Aug. 1755, NYHS *Collections*, 54 (1921): 23–24.

27. C. B. Wilde, "Hutchinsonianism, Natural Philosophy and Religious Controversy in Eighteenth Century Britain," *History of Science* 18 (1980): 1–24; for Hutchinsonian aether: Brooke, "Why Did the English Mix Their Science and Their Religion?" 71.

28. Samuel Pike to C. Colden, 10 July 1753, NYHS *Collections,* 53 (1920): 396–399; C. Colden to S. Pike, "in answer to July 1753," Colden Scientific Manuscripts, microfilm, NYHS; C. Colden to S. Pike, 12 May 1755, NYHS *Collections*, 54 (1921): 7–9; C. Colden to S. Pike, June 1755, Colden Scientific Manuscripts; for "not increased my vanity much": C. Colden to B. Franklin, 19 Nov. 1753, NYHS *Collections*, 53 (1920): 413–418; for "very short": Colden, *Principles of Action*, 73; for Pike: Schofield, *Mechanism and Materialism*, 124–125.

29. Advertisement (with variants): *NYEP*, 2, 9, and 16 Nov. 1747, 7, 14, 21, and 28 Dec. 1747; for " bedawbing": *NYEP*, 8 Feb. 1748; for "profound Conjuror":

NYEP, 16 Nov. 1747; "An Address to the Freeholders and Freeman," *NYWPB*, 18 Jan. 1748; other such items: *NYEP*, 7 Dec. 1747; *NYWPB*, 25 Jan. 1748; *NYEP*, 1 Feb. 1748; *NYEP*, 1 Feb. 1748; *NYEP*, 15 Feb. 1748; *NYEP*, 22 Feb. 1748.

30. For sending parts of the revision: C. Colden to P. Collinson, [Oct. 1755], NYHS *Collections*, 54 (1921): 38–39; for poor sales of 1751 edition: P. Collinson to C. Colden, 6 March 1759, NYHS *Collections*, 54 (1921): 296–298; for offer to keep it in library: R. Whytt to C. Colden, 16 May 1763, NYHS *Collections*, 55 (1922): 217–219.

31. Ad for King's College that described curriculum: *NYM*, 3 June 1754; David C. Humphrey, *From King's College to Columbia*, 106, 123, 168–169; mathematical school example: ad of Thomas Carroll, *NYM*, 13 May 1765; for "at the Sign of Newton's Head": *NYM*, 1 Dec. 1760; for lecture advertisements: D. Eccleston, *NYWPB*, 15 Oct. 1770; C. Colles, *RNYG*, 24 March 1774; Bryerly and Day, *RNYG*, 22 June 1774; Abraham Weatherwise, Gent., *Father Abraham's Almanack . . . 1759* (New-York: Hugh Gaine, "for 1759").

32. Harry Woolf, *The Transits of Venus: A Study of Eighteenth-Century Science* (Princeton: Princeton University Press, 1959).

33. For attempt to coax Colden to participate, see J. Alexander to C. Colden, 2 Feb. 1753, NYHS *Collections*, 53 (1920): 368–369; James Alexander, *Letters Relating to the Transit of Mercury over the Sun, which is to happen May 6, 1753* ([Philadelphia: B. Franklin, 1753]); for day was cloudy: J. Alexander to C. Colden, 10 May 1753, NYHS *Collections*, 53 (1920): 388; for reports of the transit of Venus of 1761 elsewhere: *NYM*, 3 Aug. 1761; *NYM*, 19 Oct. 1761.

34. For preparations for watching the transit, see *NYJ*, 25 May 1769; for Skinner's reports: *NYJ*, 1 and 29 June 1769; *NYWPB*, 12 June 1769; *NYJ*, 6 July 1769; for Skinner: Gregory Palmer, *Biographical Sketches of Loyalists of the American Revolution* (Westport, CT: Meckler Publishing, 1984), 792; for Harpur's transit of Venus: *NYJ*, 29 June 1769; for Harpur observing a transit of Mercury: *NYJ*, 7 Dec. 1769.

35. For awareness of necessity of comparing observations (but no comment on absence of same in New York): *NYJ*, 10 Aug. 1769; for report of the return of Captain Cook's ship, the Endeavor, mentioning its transit of Venus activity: *NYWPB*, 23 Sept. 1771.

36. John Clapp, *An Almanack . . . 1697* ([New-York]: William Bradford, 1697); "B.A.," *An Astronomical Diary, or an Almanack . . . 1723* (New-York and Philadelphia: William Bradford and Andrew Bradford, n.d.); Titan Leeds, *The American Almanack . . . 1734* (New-York: William Bradford, 1734); for 1750s almanacs that assumed Copernicanism: Roger Sherman, *An Almanack . . . 1752* (New-York: Henry De Foreest, n.d.); Thomas More, *The American Country Almanack . . . 1755* (New-York: James Parker, n.d.); Jesse Parsons, *The American Ephemeris, or, an Almanack . . . 1757* (New-York: J. Parker and W. Weyman, n.d.).

37. Frank Freeman, *Freeman's New-York Almanack . . . 1767* ([New-York]: John Holt, n.d.)

38. For Americanior's reports, see *NYM*, 2 and 16 Feb., 2, 23, and 30 March 1767; *NYJ*, 19 March and 9 April 1767; letter from "Rhodensius": *NYJ*, 9 April 1767; for other reports of efforts to measure weather: *NYWJ*, 23 Jan. 1743/44; *NYWPB*,

27 Jan. 1752; *NYWPB*, 3 Feb. 1752; Roger More, *Poor Roger. The American Country Almanack . . . 1757* (New-York: J. Parker and W. Weyman, n.d.); *NY-WPB*, 31 Jan. 1765; *NYM*, 4 Feb. 1765; *NYJ*, 28 Feb. 1771; *NYJ*, 5 March 1772; H. Howard Frisinger, *The History of Meteorology: To 1800* (New York: Science History Publications, 1977); Peter Eisenstadt, "The Weather and Weather Forecasting in Colonial America" (PhD diss., New York University, 1990).

39. For letter comparing almanacs, see *NYM*, 23 March 1767; for Freeman's concession: Frank Freeman, *Freeman's New-York Almanack . . . 1768* ([New-York]: John Holt, n.d.).

40. For example of agricultural directions, see Roger More, *Poor Roger. The American Country Almanack . . . 1767* (New-York: James Parker, n.d.); for report of girdling trees, cutting brush: Jared Elliot, "Essay on Field Husbandry," *NYWPB*, 4 June 1753; for Borghard: David Steven Cohen, *The Dutch American Farm* (New York: New York University Press, 1992), 156; for verse: Roger More, *Poor Roger. The American Country Almanack . . . 1771* (New-York: Samuel Inslee and Anthony Car, n.d.); for persistence in British almanacs: Patrick Curry, *Prophecy and Power: Astrology in Early Modern England* (Princeton: Princeton University Press, 1989), 95–116.

41. For suppression of prediction, see Curry, *Prophecy and Power*, 45–89; for example of explanation of how eclipses happen: Titan Leeds, *The American Almanack . . . 1725* (New-York: William Bradford, n.d.); for example of woodcut illustrating a partial eclipse: William Birkett, *Birkett, 1738. An Almanack . . . 1738* (New-York: William Bradford, 1738); William Birkett, *Poor Will's Almanack . . . 1743* ([New-York]: William Bradford, 1743); for example of prediction in the 1750s: Thomas More, *The American Country Almanack . . . 1753* (New-York: James Parker, n.d.).

42. For "wise Zealots": John Nathan Hutchins, *An Almanack . . . 1753* ([New-York]: Hugh Gaine, n.d.); for reference to Leeds's prediction (which was in his 1738 almanac): John Nathan Hutchins, *Hutchin's Improved: Being an Almanack . . . 1764* (New-York: Hugh Gaine, n.d.).

43. Sara Schechner Genuth, *Comets, Popular Culture, and the Birth of Modern Cosmology* (Princeton: Princeton University Press, 1997); Simon Schaffer, "Newton's Comets and the Transformation of Astrology," *Astrology, Science and Society: Historical Essays*, ed. Patrick Curry (Woodbridge, UK: The Boydell Press, 1987).

44. J. Alexander to C. Colden, 15 Feb. 1743/44, NYHS *Collections*, 52 (1919): 48–49; [J. Alexander], "Observations on a Comet," 22 Jan. 1743/44, Colden Scientific Manuscripts, microfilm, NYHS; for essays on comets: *NYEP*, 2 Feb. 1743/44; *NYWPB*, 12 Feb. 1743/44; for report of the 1744 comet: *NYWPB*, 9 Jan. 1743/44; for poem describing it as warning from God: *NYWPB*, 19 March 1743/44; Parsons, *American Ephemeris . . . 1757*; for "now be look'd for": 5 June 1758, *NYWPB*; for Halley's Comet: Roger More, *Poor Roger. The American Country Almanack . . . 1758* (New-York: James Parker and William Weyman, n.d.); *NYM*, 21 Nov. 1757; *NYWPB*, 21 Nov. 1757; *NYWPB*, 28 Nov. 1757; *NYWPB*, 5 Dec. 1757; *NYM*, 19 Dec. 1757; *NYWPB*, 16 April 1759; *NYWPB*, 14 May 1759; *NYWPB*, 28 Jan. 1760; *NYWPB*, 18 Feb. 1760; *NYWPB*, 3 March 1760.

45. For Skinner's essays, see *NYJ*, 21 and 28 Sept., 5 and 12 Oct., 2, 16, and 23 Nov. 1769; *NYWPB*, 11, 18, and 25 Sept, 19 and 30 Oct., 13 and 27 Nov. 1769; note that the third essay did not appear in the *Journal* and the fourth did not appear in the *Post-Boy.*

46. For New Jersey letter, see *NYJ*, 7 Sept. 1769; for Connecticut letter: *NYJ*, 5 Oct. 1769; for "guilty nations tremble" poem: *NYJ*, 28 Sept. 1769; for "idle and contemptible" charge: *NYJ*, 26 Oct. 1769; for reports without interpretation: *NYJ*, 7 Sept., 2 Nov., 23 Nov. (two reports), and 30 Nov. 1769; for assertions that comets are "Not Works of Wonder more than Works of Love": poem, *NYJ*, 5 Oct. 1769; Hugh Williamson, "A Dissertation on Comets," *The American Magazine, or General Repository* (Philadelphia, but advertised for sale in New York), Sept. 1769; doubt that one can know the significance of this or any other comet: Roger More, *Poor Roger. The American Country Almanack . . . 1770* (New-York: James Parker).

47. For "PRODIGIES," see first Skinner essay, *NYWPB*, 11 Sept. 1769, and *NYJ* 21 Sept. 1769; for "well-rectified Matter": *NYJ*, 5 Oct. 1769.

48. William Samuel Johnson to Samuel Johnson, 21 June 1754, in Johnson, *Samuel Johnson, President of King's College*, 1: 188–190; Samuel Johnson to William Samuel Johnson, [late June], 1754, *Samuel Johnson, President of King's College*, 1: 190–192.

49. Mary Cooper, entry of Aug. 5, 1769, *The Diary of Mary Cooper: Life on a Long Island Farm 1768–1773*, ed. Field Horne (Oyster Bay, NY: Oyster Bay Historical Society, 1981), 17.

CHAPTER 7   *Landscape Reimagined*

1. For geography and cartography, see David N. Livingstone and Charles W. J. Withers, eds., *Geography and Enlightenment* (Chicago: University of Chicago Press, 1999); Anne Marie Claire Godlewska, *Geography Unbound: French Geographic Science from Cassini to Humbolt* (Chicago: University of Chicago Press, 1999), 21–55; Bernard W. Heise, "Visions of the World: Geography and Maps During the Baroque Age, 1550–1750" (PhD diss., Cornell University, 1998); for racial idioms: Roxann Wheeler, *The Complexion of Race: Categories of Difference in Eighteenth-Century British Culture* (Philadelphia: University of Pennsylvania Press, 2000); articles in the special issue "Constructing Race," *William and Mary Quarterly*, 3rd series, 54 (January 1997).

2. For the Iroquois deed, see Lt. Governor Nanfan to the Lords of Trade, 20 Aug. 1701, in *DRCHSNY* IV: 888–889; Daniel K. Richter, *The Ordeal of the Longhouse: The Peoples of the Iroquois League in the Era of European Colonization* (Chapel Hill: Omohundro Institute for Early American History and Culture by University of North Carolina Press, 1992), 190–213.

3. For the Iroquois in London: Eric Hinderaker, "The 'Four Indian Kings' and the Imaginative Construction of the First British Empire," *William and Mary Quarterly*, 3rd series, 53 (1996): 487–526; Board of Trade, "State of the British Plantations in America in 1721," in *DRCHSNY* V: 591–634.

4. For the watershed argument, see Max Savelle, with Margaret Anne Fisher, *The Origins of American Diplomacy: The International History of Angloamerica,*

1492–1763 (New York: Macmillan, 1967), 206–208; for French state support for mapmaking: Josef W. Konvitz, *Cartography in France, 1660–1848: Science, Engineering, and Statecraft* (Chicago: University of Chicago Press, 1987); David Buisseret, *Mapping the French Empire in North America* (Chicago: University of Chicago Press, 1991.)

5. *Papers relating to an act of the Assembly of the Province of New-York, for encouragement of the Indian Trade, etc., and for prohibiting the Selling of Indian Goods to the French, viz., of Canada. Published by Authority* (New York: William Bradford, 1724); Cadwallader Colden, *The History of the Five Indian Nations Depending on the Province of New-York in America* (1727; reprint, Ithaca, NY: Cornell University Press, 1958), quotes on vi, xxi.

6. For King William's War, Queen Anne's War, and New Yorkers' indifference, see Michael Kammen, *Colonial New York: A History* (Millwood, NY: KTO Press, 1975), 143–145, 193–196; for Burnet's interdiction: Governor Burnet to the Lords of Trade, 26 Nov. 1720, in *DRCHSNY* V: 576–580; Thomas Elliot Norton, *The Fur Trade in Colonial New York, 1686–1776* (Madison: University of Wisconsin Press, 1974), 56–64.

7. *Papers Relating to an Act of the Assembly*, 8; Governor Burnet to the Lords of Trade, 7 Nov. 1724, in *DRCHSNY* V: 711–713.

8. [John Oldmixon], *The British Empire in America*, 2 vols. (London: John Nicholson, et al., 1708), I: 117–133, with Long Island as north on p.118; repeated error: [John Oldmixon], *The British Empire in America*, 2nd ed. (London: J. Brotherton, et al., 1741), I: 237; Herman Moll, *The Compleat Geographer: Or, the Chorography and Topography Of all the Known Parts of the Earth*, 4th ed. (London: J. Knapton, et al., 1723), 202–203.

9. For poorly supported garrison soldiers: H. V. Bowen, *Elites, Enterprise, and the Making of the British Overseas Enterprise, 1688–1775* (Houndsmills, UK: Macmillan Press, 1996), 23–25; for modest and intermittent government interest in the North American colonies: Jacob M. Price, "Who Cared About the Colonies? The Impact of the Thirteen Colonies on British Society and Politics, 1714–1775," in *Strangers in the Realm: Cultural Margins of the First British Empire*, ed. Bernard Bailyn and Philip D. Morgan (Chapel Hill: OIEAHC/University of North Carolina Press, 1991).

10. Daniel Leeds, *The American Almanac for the Year of Christian Account 1706* (New-York: William Bradford, 1706); for Turkish empire: *NYG*, 22 April and 13 May 1728; for Guinea trade: *NYG*, 20 July 1730; for Confucius: *NYWJ*, 24 and 30 Jan. 1737/38, 6 and 13 Feb. 1737/38; for reading Hennepin: Joseph Morgan to the Royal Society, 22 Dec. 1732, in Royal Society of London, "Letters and Communications from Americans, 1662–1900," microfilm, APS, reel 3.

11. First Bradford advertisement of maps: Daniel Leeds, *An Almanack . . . 1695* (New-York: William Bradford, 1694); for quoted advertisement: *NYG*, 26 Nov. 1733.

12. For latitude: Charles Wooley, "A Two Years Journal in New York and Part of Its Territories in America" (1701), reprinted in Cornell Jaray, ed., *Historic Chronicles of New Amsterdam, Colonial New York, and Early Long Island* (Port Washington, NY: Ira J. Friedman, n.d.), 22; for longitude: William Burnet,

"Observations on the Eclipses of the first Satellite of Jupiter, communicated by his Excellency William Burnett, Esq; Governor of New York, F.R.S." in Royal Society of London, *Philosophical Transactions* 33 (1724–1725): 162–164; for correction of longitude: the Rev. Mr. James Bradley, "The Longitude of London, and the Fort of New York, from Wansted and London, determin'd by Eclipses of the First Satellite of Jupiter," Royal Society of London, *Philosophical Transactions* 34 (1726–1727): 85–90.

13. For example of asking directions, Phineas Stevens, "Journal of Captain Phineas Stevens' Journey to Canada, 1752," in Newton D. Mereness, ed., *Travels in the American Colonies* (New York: The Macmillan Co., 1916), 305; for the *Vade Mecum*: Carl Bridenbaugh, "Introduction" to Alexander Hamilton, *Gentleman's Progress: The Itinerarium of Dr. Alexander Hamilton, 1744*, ed. Carl Bridenbaugh (Westport, CT: Greenwood Press, 1948), xv.; for Lynn map: Edwin G. Burrows and Mike Wallace, *Gotham: A History of New York City to 1898* (New York: Oxford, 1999), 139; for Lynn map advertised: *NYG*, 30 Aug. 1731.

14. Hamilton, *Gentleman's Progress*, 60, 91, 172; for canoes, elm bark: Richard Smith, *A Tour of Four Great Rivers, the Hudson, Mohawk, Susquehanna, and Delaware in 1769*, ed. Francis W. Halsey (New York: Charles Scribner's Sons, 1906), 10–11; for raccoon: Henry Melchior Muhlenberg, *The Journals of Henry Melchior Muhlenberg*, trans. Theodore G. Tappert and John W. Doberstein, 3 vols. (Philadelphia: The Muhlenberg Press, 1942), I: 248; for baby raccoons: Cadwallader Colden to Peter Collinson, NYHS *Collections*, 51 (1918): 278; for purchases: Ann (MacVickar) Grant, *Memoirs of an American Lady* (1901; reprint, Freeport, NY: Books for Libraries Press, 1972), 131–133; for corn seizure: Shirley W. Dunn, *The Mohicans and Their Land, 1609–1730* (Fleischmanns, NY: Purple Mountain Press, 1994), 161.

15. For reports of slaves committing crimes in the 1730s, see *NYG*, 14 Aug. 1732; *NYG*, 28 Jan. 1733/34; *NYG*, 25 July 1736; *NYG*, 18 July 1737; *NYG*, 12 Sept. 1737; for reports of slave conspiracies in the 1730s: *NYG*, 2 Nov. 1730; *NYG*, 30 Nov. 1730; *NYG*, 21 Dec. 1730; *NYG*, 3 May 1731; *NYG*, 28 Jan. 1733/34; *NYG*, 18 March 1733/34; *NYG*, 16 April 1734; *NYWJ*, 29 July 1734; *NYG*, 27 April 1734; *NYWJ*, 29 July 1734; *NYG*, 5 Aug. 1734; *NYG*, 25 Nov. 1734; *NYG*, 1 March 1734/35; *NYG*, 22 Nov. 1736; *NYG*, 29 Nov. 1736; *NYG*, 22 Feb. 1736/37; *NYG*, 22 March 1736/37; *NYG*, 7 May 1739; Anon., *A Scheme (By Striking Twenty Thousand Pounds, Paper Money) To Encourage the Raising of Hemp, and The Manufacturing of Iron in the Province of New-York* (New York: n.p., 1737).

16. Daniel Horsmanden, *A Journal of the Proceedings in the Detection of the Conspiracy formed by Some White People, in Conjunction with Negro and other Slaves, for Burning the City of New-York in America, And Murdering the Inhabitants*, ed. and with an introduction by Thomas J. Davis (1744; reprint, Boston: Beacon Press, 1971); see also Serena R. Zabin, "Places of Exchange: New York City, 1700–1763" (PhD diss., Rutgers University, 2000), 137–173; Charles Peter Hoffer, *The Great New York Conspiracy of 1741: Slavery, Crime, and Colonial Law* (Lawrence: University Press of Kansas, 2003); Jill Lepore, *New York Burning: Liberty, Slavery, and Conspiracy in Eighteenth-Century Manhattan* (New York: Alfred A. Knopf, 2005).

17. For the trial of John Ury, see Horsmanden, _Journal of the Proceedings_, 332–375; for Ury's execution speech: 379–381; for Protestant-Catholic struggle: 369–371; for five Spanish men: 260–262; for Oglethorpe's letter: 350–351; for Spanish war plot: 341–342; for "the most abandoned whites": 419.

18. For Horsmanden quotes, see _Journal of the Proceedings_, 107, 109–110; for Smith quotes: _Journal of the Proceedings_, 105–106.

19. For undated, unsigned letter received by Elizabeth (Colden) DeLancey, see NYHS _Collections,_ 67 (1934): 269–272.

20. For Kerry's child, see Horsmanden, _Journal of the Proceedings_, 15; for black women delivering twins of different colors: _NYWJ_, 27 Jan. 1734/35; _NYWPB_, 13 March 1766; for Hughson and Caesar: _Journal of the Proceedings_, 273–276.

21. For privateers, see Kammen, _Colonial New York_, 330; for examples of news items and essays with geographic descriptions relevant to the war: _NYWPB_, 4 Nov. 1745; _NYWPB_, 30 June 1746; _NYWPB_, 11 August 1746; _NYWPB_, 20 April 1747; for advertisements for plans of Louisbourg: _NYWPB_, 10 June 1745, 29 Sept. 1746.

22. For King George's War, see Kammen, _Colonial New York_, 298, 305–307, 311.

23. For Mohawks scalping, see _NYEP_, 28 July 1746; for French-allied Indians scalping: _NYEP_, 24 Aug. 1746; for drinking blood: _NYEP_, 5 Aug. 1745; Cadwallader Colden to Gronovius, 30 May 1746, "Selections from the Scientific Correspondence of Cadwallader Colden," arranged by Asa Gray, _American Journal of Science and Arts_ 44 (1843): 101; for a similar association of "barborous" actions: Governor George Clinton to the Lords of Trade, 27 March 1745, in _DRCHSNY_ VI: 281–282.

24. Thomas More, _The American Country Almanack . . . 1746_ (New York: James Parker, 1745); for reports of slave conspiracies and rebellions elsewhere: _NYWPB_, 21 Feb. 1743/44; _NYG_, 24 Sept. 1744; _NYWPB_, 8 April 1745; _NYEP_, 8 April 1745; _NYEP_, 18 Nov. 1745; _NYWPB_, 13 Jan. 1745/46; _NYEP_, 10 March 1745/46; _NYEP_, 3 April 1749; Irish servants: _NYWPB_, 24 Dec. 1744.

25. For Duyckinck's ad, see _NYWJ_, 16 Oct. 1749; for ad for geography books: _NYWPB_, 12 Sept. 1748; for curriculum: Cadwallader Colden to Samuel Johnson, 20 Dec. 1752, in _Samuel Johnson, President of King's College: His Career and Writings_, ed. Herbert Schneider and Carol Schneider, 4 vols. (New York: Columbia University Press, 1929), 2: 300–301; for boundary negotiations: _NYEP_, 13 July 1752; _NYWPB_, 4 March 1754; for Northwest Passage: _NYWPB_, 30 April 1750; _NYWPB_, 15 Jan. 1753; for discoveries in the South Seas: _NYWPB_, 26 Feb. 1753.

26. For ad for primmer, see _NYEP_, 9 July 1750; for essays on New York: Milton M. Klein, ed., _The Independent Reflector, or, Weekly Essays on Sundry Subjects More particularly adapted to the Province of New-York_ (Cambridge: Harvard University Press, 1963), 103–108, 433–438; Thomas More, _The American Country Almanack . . . 1751_ (New-York: James Parker, n.d.).

27. For "your Yorck Gentlemen," see David Schuyler, Peter Schuyler, and Nicholas Pecherd to Mr. "Goolding" (_sic_, for Cadwallader Colden), 25 Oct. 1753, NYHS _Collections_, 53 (1920): 412; for example of Dutch awareness of English ridicule: letter of Johanes Ongewardeerd, _NYEP_, 16 Sept. 1751; for comment that Dutch magistrates show partiality toward Dutch: William Johnson to Goldsbrow Banyar,

10 Feb. 1761, in *The Papers of Sir William Johnson*, ed. James Sullivan, Milton Hamilton, and Albert B Corey, 13 vols. (Albany: University of the State of New York, 1921–1962), III: 326–329; for Classis of Amsterdam: *ERSNY*, V: 3652; for pamphlet on Louisbourg: *NYWPB*, 2 February 1746/47; for "cries of the oppressed": *NYWJ*, 27 June 1748.

28. For Evans's map first advertised for sale, see *NYWPB*, 2 Oct. 1749; for Evans: Walter Klinefelter, "Lewis Evans and His Maps," American Philosophical Society, *Transactions* 61 (1971): 3–65; Lawrence Henry Gipson, *Lewis Evans* (Philadelphia: The Historical Society of Pennsylvania, 1939); Margaret Beck Pritchard and Henry G. Taliaferro, *Degrees of Latitude: Mapping Colonial America* (Williamsburg, VA: Colonial Williamsburg Foundation in association with Harry N. Abrams, 2002), 150–151, 172.

29. For A.B.'s letter, see *NYWPB*, 1 May 1749; for Evans's response: *NYWPB*, 15 May 1749; for announcement that revised edition was published with "a Degree of Correctness": *NYWPB*, 27 Aug. 1753.

30. For news of map, see *NYWPB*, 5 May 1755; *NYWPB*, 21 July 1755; for advertisements: *NYWPB*, 25 August 1755 and six times thereafter; *NYM*, 1 Sept. 1755 and four times thereafter; for comment on "Beaver Hunting" grounds: Goldsbrow Banyar to William Johnson, 15 Aug. 1755, *Papers of Sir William Johnson*, I: 850; Henry N. Stevens, *Lewis Evans: His Map of the Middle British Colonies in America* (1920; reprint, n.p.: Arno Press and the New York Times, 1970).

31. Lewis Evans, *Geographical, Historical, Political, Philosophical, and Mechanical Essays* (Philadelphia: B. Franklin and D. Hall, 1755).

32. Lawrence Henry Gipson, *The Great War for Empire: The Years of Defeat, 1754–1757*, vol. 6 of *The British Empire before the American Revolution*, 9 vols. (New York: Alfred A. Knopf, 1946), 58–60, 73–74.

33. James Thomas Flexner, *Mohawk Baronet: Sir William Johnson of New York* (Syracuse: Syracuse University Press, 1959), 126–129; William Smith Jr., *The History of the Province of New-York, from Its Discovery to the Appointment of Governor Colden, in 1762* (New York: NYHS, 1829), I: 228.

34. For letters of Roger Mortar, see *NYM*, 5 Jan. and 2 Feb. 1756. The January 5 letter was reprinted in a pamphlet with a rebuttal from Evans: Lewis Evans, *Geographical, Historical, Political, Philosophical and Mechanical Essays. Number II* (Philadelphia: Lewis Evans; New-York: Garrett Noel, 1756).

35. For the war in New York, see Kammen, *Colonial New York*, 314–330.

36. For newspaper accounts with descriptions of geography: *NYWPB*, 17 Feb. 1755; *NYM*, 19 Jan. 1756; *NYWPB*, 1 and 29 May 1758; *NYWPB*, 5 June 1758; *NYWPB*, 7 Jan. 1760; *NYM*, 18 Aug. 1760; for the war in almanacs: Roger More, *Poor Roger. The American Country Almanack . . . 1757* (New-York: J. Parker and W. Weyman, n.d.); John Nathan Hutchins, *An Almanack . . . 1759* (New-York: Hugh Gaine, n.d.); Abraham Weatherwise, *Father Abraham's Almanack . . . 1759* (New York: James Parker, n.d.); Thomas More, *Poor Thomas Improved: Being More's Country Almanack . . . 1760* (New-York: William Weyman, n.d.); for advertisements for books and pamphlets: *NYM*, 1 Sept. 1755; *NYM*, 14 June 1756; *NYM*, 3 Nov. 1755 and seven subsequent issues; *New American Magazine* (Woodbridge, NJ: James Parker), Jan.–Aug., 1758; *NYM*, 29 Oct. 1759; *NYM*, 10

Nov. 1760; *NYM*, 20 July 1761; *NYWPB*, 20 Aug. 1761; for advice to familiarize oneself with geography: *NYWPB*, 2 July 1758.

37. For advertisement for Blodget's map, see *NYM*, 26 Jan.1756, with eleven more through May 1756; for Louisbourg: *NYM*, 14 Aug. 1758, and *NYWPB*, 14 Aug. 1758; for Quebec: More, *Poor Thomas Improved . . . 1760* ; for a plan of Niagara drawn by George Demler of the Sixtieth Regiment and engraved by Michael DeBruls: *NYWPB*, 15 April 1762; for other new maps: Atlantic coast, *NYM*, 27 March 1758; Nova Scotia, *NYWPB*, 3 Dec. 1759; Pennsylvania, *NYM*, 12 March 1759; St. Lawrence River, *NYM*, 22 Sept. 1760; for updated Evans map: *NYM*, 23 June 1760; for the Royal Engineers: James Montresor, "Journals of Colonel James Montresor, 1757–1759," NYHS *Collections*, 14 (1881): 16–58; John Montressor, "Journals of Captain John Montresor," NYHS *Collections*, 14 (1881): 115–420; Bruce Robertson, "Venit, Vidit, Depinxit: The Military Artist in America," in *Views and Visions: American Landscape Before 1830*, ed. Edward J. Nygren with Bruce Robertson (Washington: Corcoran Gallery of Art, 1986); Douglas W. Marshall, "The British Engineers in America, 1755–1783," *Journal of the Society for Army Historical Research* 51 (1973): 155–163; W. P. Cumming, "The Montresor-Ratzer-Sauthier Sequence of Maps of New York City, 1766–1776," *Imago Mundi* 31 (1979): 55–65; for "till very lately, we knew little of": *NYWPB*, 10 Sept. 1761.

38. For headline, see *NYWPB*, 18 Sept. 1760; for "restless . . . Enemy": John Nathan Hutchins, *Hutchins's Almanack . . . 1759* (New-York: Hugh Gaine, n.d.); for header poems: Roger More, *Poor Roger. The American Country Almanack . . . 1761* (New-York: James Parker, n.d.); for "the Triumphs of our Nation": Thomas More, *Poor Thomas Improved: Being More's Country Almanack . . . 1761* (New-York: W. Weyman, n.d.).

39. For first advertisement for the *American Gazeteer*, see *NYM*, 25 Jan. 1762; William Smith, *History of the Province of New-York, From the First Discovery to the Year 1732* (London: Thomas Wilcox, 1757); for sample advertisement for Smith: *NYM*, 6 March 1758; for sample advertisement for local views: *NYM*, 16 Nov. 1761; Sir William Johnson, for one, evidently purchased several of these views: Michael DeBruls to William Johnson, 11 Jan. 1763, *Papers of Sir William Johnson*, IV: 12; Gloria Gilda Deák, *Picturing America, 1497–1899: Prints, Maps, and Drawings Bearing on the New World Discoveries and on the Development of the Territory that is now the United States* (Princeton: Princeton University Press, 1988), 66–68; advertisement for Kalm: *NYWPB*, 12 Oct. 1772.

40. Philip J. Schwartz, *The Jarring Interests: New York's Boundary Makers, 1664–1776* (Albany: State University of New York Press, 1979), 184, 219; Ray Billington, "The Fort Stanwix Treaty of 1768," *New York History* 25 (1944): 182–194; Michael A. Bellesiles, *Revolutionary Outlaws: Ethan Allen and the Struggle for Independence on the Early American Frontier* (Charlottesville: University Press of Virginia, 1993), 31.

41. Evarts B. Greene and Virginia D. Harrington, *American Population Before the Federal Census of 1790* (New York: Columbia University Press, 1932), 88–91; Aaron Fogleman, "Migrations to the Thirteen British North American Colonies, 1700–1775: New Estimates," *Journal of Interdisciplinary History* 22 (1992):

691–709; for "Land Mad": George Croghan to William Johnson, 30 March 1766, *Papers of Sir William Johnson*, V: 128–130; Wyllys Terry, "Negotiating the Frontier: Land Patenting in Colonial New York" (PhD diss., Boston University, 1997), 196–231.

42. Roger More, *Poor Roger. The American Country Almanack . . . 1758* (New-York: James Parker and William Weyman, n.d.); for "Tortures and unprecedented Cruelties": *New American Magazine*, Jan. 1758; for "feroce Bipeds": *RNYG*, 2 Sept. 1773; Peter Middleton, *A Medical Discourse, or an Historical Inquiry into the Ancient and Present State of Medicine* (New-York: Hugh Gaine, 1769), 48–49; for without agriculture: William Smith, *The History of the Province of New-York* (London: Thomas Wilcox, 1757), 52–54.

43. For order to stop granting land, see 23 Nov. 1761, in *DRCHSNY* VII: 472–476; for order to vacate Kayaderosseras Patent: Lords of Trade to Cadwallader Colden, 10 July 1764, in *DRCHSNY* VII: 633–634; for Assembly's refusal: William Johnson to the Lords of Trade, 30 Oct. 1764, in *DRCHSNY* VII: 670–675; for "unlettered Barbarians": *NYM*, 22 Oct. 1764; for claim that fraud against Indians was impossible in New York: Cadwallader Colden to the Lords of Trade, 12 Oct. 1764, in *DRCHSNY* VII: 667–670.

44. "Report of the Lords of Trade on the Petition of the Wappinger Indians," 30 Aug. 1766, in *DRCHSNY* VII: 868–870; William Johnson to the Earl of Shelburne, 15 Jan. 1767, in *DRCHSNY* VII: 891–894; for "mere mob": Cadwallader Colden to the Earl of Halifax, 13 Feb. 1764, in *DRCHSNY* VII: 609–610.

45. For Assembly: *NYM*, 22 Oct. 1764; for "Rights and Privileges": Governor Moore to the Earl of Shelburne, 3 April 1767, in *DRCHSNY* VII: 915–916; for Tryon's claim: Governor Tryon to the Earl of Dartmouth, 2 June 1773, in *DRCHSNY* VIII: 373–376; for the Wappingers: Georgiana C. Nammack, *Fraud, Politics, and the Dispossession of the Indians: The Iroquois Land Frontier in the Colonial Period* (Norman: University of Oklahoma Press, 1969), 70–72; Patricia U. Bonomi, *A Factious People: Politics and Society in Colonial New York* (New York: Columbia University Press, 1971), 219–224.

46. Paul David Nelson, *William Tryon and the Course of Empire: A Life in the British Imperial Service* (Chapel Hill: University of North Carolina Press, 1990), 101–109; for Sauthier: William Cumming, *The Southeast in Early Maps*, 3rd ed., rev. by Louis DeVorsey (Chapel Hill: University of North Carolina Press, 1998), 32; for Green Mountains: Bellesiles, *Revolutionary Outlaws*, 27–32, 70–111.

47. William Tryon to Lord Dartmouth, 5 Jan. 1773, in *DRCHSNY* VIII: 342–346; Dartmouth to Tryon, 3 March 1773, in *DRCHSNY* VIII: 356–357; Tryon to Dartmouth, 11 April 1772, in *DRCHSNY* VIII: 293–294; "Report of Governor Tryon on the Province of New York," in *DRCHSNY* VIII: 434–457.

48. For example of literate slave, see *NYM*, 26 April 1762; for skills, see ads in Graham Russell Hodges and Alan Edward Brown, eds., *"Pretends to be free": Runaway Slave Advertisements from Colonial and Revolutionary New York and New Jersey* (New York: Garland Publishing Company, 1994); Michael E. Groth, "Laboring for Freedom in Dutchess County," in *Mighty Change, Tall Within: Black Identity in the Hudson Valley*, ed. Myra B. Young Armstead (Albany: State University of New York Press, 2003); for shopkeepers' trade: Aileen B. Agnew,

"Living in a Material World: African Americans and Economic Identity in Colonial Albany," in *Mighty Change, Tall Within*; see also John Wood Sweet, *Bodies Politic: Negotiating Race in the American North, 1730–1830* (Baltimore: Johns Hopkins University Press, 2003).

49. Jupiter Hammon, *America's First Negro Poet: The Complete Works of Jupiter Hammon of Long Island*, ed. Stanley Austin Ransom Jr. (Port Washington, NY: Kennikat Press, 1970); Samson Occum, "An Account of the Montauk Indians, on Long Island, A.D. 1761," Massachusetts Historical Society, *Collections* 10 (1809): 106–110; John A. Strong, *The Montaukett Indians of Eastern Long Island* (Syracuse, NY: Syracuse University Press, 2001), 63–79; Jean Fitz Hankins, "Bringing the Good News: Protestant Missionaries to the Indians of New England and New York, 1700–1775" (PhD diss., University of Connecticut, 1993); news report of a Mohegan preacher: *NYJ*, 22 Dec. 1774; Graham Russell Hodges, *Root and Branch: African Americans in New York and East Jersey, 1613–1863* (Chapel Hill: University of North Carolina Press, 1999), 119–126; John Ogilvie, "Diary, 1750–1759," ed. Milton W. Hamilton, *Bulletin of the Ticonderoga Museum* 10 (1961): 331–385; Mary Cooper, *The Diary of Mary Cooper: Life on a Long Island Farm, 1768–1773*, ed. Field Horne (Oyster Bay, NY: Oyster Bay Historical Society, 1981), 13, 18, 23–24.

INDEX

Page numbers in italics refer to illustrations.

About the Author

Sara Stidstone Gronim is an assistant professor of history at the C. W. Post campus of Long Island University, where she teaches courses on colonial and Revolutionary history, and the history of medicine and the environment. She received her PhD in early American history from Rutgers University in 1999. She lives with her husband in Park Slope, Brooklyn, and has two grown children. She is the author of a number of articles about the relationship between people and the natural world. This is her first book.